Novel Techniques for Dialectal Arabic Speech Recognition

T0191989

Mohamed Elmahdy · Rainer Gruhn ·
Wolfgang Minker

Novel Techniques for Dialectal Arabic Speech Recognition

 Springer

Mohamed Elmahdy
Qatar University
Doha
Qatar

Rainer Gruhn
SVOX Deutschland GmbH
Ulm
Germany

Wolfgang Minker
Institute of Information Technology
University of Ulm
Ulm
Germany

ISBN 978-1-4899-9945-0   ISBN 978-1-4614-1906-8 (eBook)
DOI 10.1007/978-1-4614-1906-8
Springer New York Dordrecht Heidelberg London

Printed on acid-free paper

Springer is part of Springer Science+Business Media (www.springer.com)

For my parents, wife, and kids

Preface

This book describes novel approaches to improve automatic speech recognition for dialectal Arabic. Since the existing dialectal Arabic speech resources, that are available for the task of training speech recognition systems, are very sparse and are lacking quality, we describe how existing Modern Standard Arabic (MSA) speech resources can be applied to dialectal Arabic speech recognition. Our assumption is that MSA is always a second language for all Arabic speakers, and in most cases we can identify the original dialect of a speaker even though he is speaking MSA. Hence, an acoustic model trained with a sufficient number of MSA speakers from different origins will implicitly model the acoustic features for the different Arabic dialects and in this case, it can be called dialect-independent acoustic modeling.

In this work, Egyptian Colloquial Arabic (ECA) has been chosen as a typical Arabic dialect. ECA is the first ranked Arabic dialect in terms of number of speakers. A high quality ECA speech corpus with accurate phonetic transcriptions has been collected. MSA acoustic models were trained using news broadcast speech data. In fact, MSA and the different Arabic dialects do not share the same phoneme set. Therefore, in order to cross-lingually use MSA in dialectal Arabic speech recognition, we propose phoneme sets normalization. We have normalized the phoneme sets for MSA and ECA. After phoneme sets normalization, we have applied state-of-the-art acoustic model adaptation techniques like Maximum Likelihood Linear Regression (MLLR) and Maximum A-Posteriori (MAP) to adapt existing phonemic MSA acoustic models with a small amount of ECA speech data. Speech recognition results indicated a significant increase in recognition accuracy compared to a baseline model trained with only ECA data. Best results were obtained when combining MLLR and MAP.

Since for dialectal Arabic in general, it is hard to phonetically transcribe large amounts of speech data, we have studied the use of grapheme-based acoustic models where the phonetic transcription is approximated to be the word letters rather than the exact phonemes sequence. A large number of Gaussians in the Gaussian mixture model is used to implicitly model missing vowels. Since dialectal Arabic is mainly spoken and not formally written, the graphemic form usually does not match the actual spelling as in MSA. A graphemic MSA acoustic model was therefore used to force align and to choose the correct ECA spelling from a set of automatically generated spelling variants lexicon. Afterwards, MLLR and MAP acoustic model adaptations are applied to adapt the graphemic MSA acoustic model with a small amount of dialectal data. In the case of graphemic adaptation, we are still able to observe a significant improvement in speech recognition accuracy compared to the baseline. In order to prove a consistent improvement across different Arabic dialects, the presented approach was re-applied on Levantine Colloquial Arabic (LCA).

Finally, we have also shown how to use the Arabic Chat Alphabet (ACA) to transcribe dialectal speech data. ACA transcriptions include short vowels that are omitted in traditional Arabic orthography. So our assumption was that ACA transcriptions are closer to the actual phonetic transcription rather than graphemic transcriptions. ACA-based and phoneme-based acoustic models performed similar but significantly outperformed the grapheme-based acoustic models.

Acknowledgements

الْحَمْدُ لِلَّهِ رَبِّ الْعَلَمِين

Praise be to Allah, the Cherisher and Sustainer of the worlds

First and foremost I would like to offer my sincerest gratitude to Prof. Slim Adbdennadher, who was always supporting me throughout this research with his patience, knowledge, guidance, trust, and encouragement. I owe my deepest gratitude to all the members of the dialogue systems group at Ulm University who were always reviewing my work and providing me with their very helpful feedbacks. Lastly, I wish to offer my regards and blessings to my parents and to my wife who were always praying for me all the time and wishing me all the success. May God bless them all.

Contents

Nomenclature

ACA   Arabic Chat Alphabet
ACAST   Arabic Chat Alphabet for Speech Transcription
ASR   Automatic Speech Recognition
CD   Context Dependent
CI   Context Independent
DTC   Distinct Tri-phone coverage
ECA   Egyptian Colloquial Arabic
G2P   Grapheme to Phoneme
GMM   Gaussian Mixture Model
HCI   Human-Computer Interaction
HMM   Hidden Markov Model
IPA   International Phonetic Alphabet
kHz   Kilo Hertz
LCA   Levantine Colloquial Arabic
MAP   Maximum A-Posteriori
MFCC   Mel Frequency Cepstral Coefficient
MLLR   Maximum Likeihood Linear Regression
MSA   Modern Standard Arabic
OOV   Out of Vocabulary
PCM   Pulse Code Modulation
PER   Phoneme Error Rate
RTF   Real Time Factor
SAMPA   Speech Assessment Methods Phonetic Alphabet
SNR   Signal to Noise Ratio
SVO   Subject Verb Object
THD   Total Harmonic Distortion
TTS   Text-to-Speech
USB   Universal Serial Bus
VSO   Verb Subject Object
WER   Word Error Rate

List of Figures

List of Tables

Chapter 1
Introduction

The goal of *Automatic Speech Recognition* (ASR) systems is to map an acoustic speech signal to a string of words. One of the major applications of ASR is in *Human-Computer Interaction* (HCI). In many HCI tasks, speech input is more appropriate than keyboard or other pointing devices where natural language is helpful or where other interfacing devices (keyboard, mouse, etc) are not appropriate. This includes telephony applications and hands-busy or eyes-busy applications.

Though Arabic language is widely spoken, research done in the area of ASR for Arabic is very limited compared to other same rank languages e.g. Chinese Mandarin. That is why, to date, the majority of HCI applications lacks the support for Arabic.

The majority of previous work in Arabic ASR has focused on the formal standard Arabic language that is known as *Modern Standard Arabic* (MSA). MSA is not the language of ordinary communications in all Arabic countries. Other Arabic varieties used in everyday life are known as the *Arabic dialects*.

Practically, MSA is acceptable for speech synthesis (text-to-speech) in any Arabic spoken language system. This is due to the fact that MSA is always a second language for all Arabic speakers. However, the real problem is usually in ASR where MSA is not used by major parts of the population.

A significant problem in Arabic ASR is the existence of quite many different dialects e.g. Egyptian, Levantine, Saudi, Iraqi, etc. Every country has its own dialect, and sometimes there exist different dialects within the same country. Moreover, the different Arabic dialects are only spoken and not formally written and significant phonological, morphological, syntactic, and lexical differences exist between the dialects and the standard form. This situation is called *diglossia* and it has been documented in (Ferguson 1959; Kaye 1970).

In order to train a reliable speaker-independent ASR system for a certain language, a high quality speech corpus should be available for that particular language. This speech corpus should be phonetically transcribed in order to train statistical models for the different phonemes in that language. In other words, the exact phoneme sequence for all speech segments in the corpus should be transcribed. A large number of different speakers (more than 200) is also required in order to

M. Elmahdy et al., *Novel Techniques for Dialectal Arabic Speech Recognition*, DOI 10.1007/978-1-4614-1906-8_1, © Springer Science+Business Media New York 2012

model different accents and hence creating a speaker independent speech recognition system (Rabiner 1989).

For dialectal Arabic, there exists very limited speech corpora for the purpose of training ASR systems. For many Arabic dialects, no speech resources exist at all. Thus, training a reliable ASR system is not easily achieved for dialectal Arabic. The development of large dialectal Arabic speech corpora remains rather difficult. This is mainly due to the diglossic nature of dialectal Arabic and due to the difficulties of estimating the correct phonetic transcription for the different Arabic dialects.

On the other hand, for MSA, there exist a lot of speech resources and many ASR approaches have been proposed to date. There are significant differences between MSA and dialectal Arabic to the extent to consider them as completely different languages. In order to improve ASR for dialectal Arabic, previous work has mainly focused on dialectal Arabic speech data collection. However, dialectal data amounts are still far away from MSA speech resources. Moreover, the majority of previous researches for dialectal Arabic ASR did not take into consideration the existing large speech resources for MSA.

Dialectal Arabic ASR problems can be summarized into two main points as follows:

1. Sparse and little speech resources are available for the purpose of training ASR systems. This is mainly due to the existence of quite many different Arabic dialects. Every country has its own dialect and sometimes there exist different dialects within the same country.
2. Phonetic transcription difficulties arise for the following reasons:
   a) Dialectal Arabic is mainly spoken and not written. Furthermore, there is no standard for dialectal Arabic orthography.
   b) There exist extra phonemes in dialectal Arabic that cannot be directly estimated neither from the script nor from diacritic marks.
   c) Phonetic transcription techniques for MSA cannot be used directly for dialectal Arabic.

In this book, our main goal is to propose novel approaches that improve ASR accuracy for dialectal Arabic. This includes finding a way to benefit from existing large MSA speech resources to improve dialectal Arabic ASR. Moreover, we propose a reliable way to phonetically transcribe dialecal Arabic speech data.

1.1 State-of-the-Art and Previous Work

The majority of previous work in Arabic ASR has focused on MSA speech recognition. Speech data are usually news broadcasts where MSA is the formal language (Maamouri et al. 2006; Yaseen et al. 2006). Different MSA-based speech recognition systems have been developed for the purpose of news broadcast transcription as in (Billa et al. 2002; Lamel et al. 2007; Rybach et al. 2007; Messauoudi 2004, 2006; Ng et al. 2009; Gales et al. 2007).

Arabic is a morphologically very rich language. That is why a simple lookup table for phonetic transcription—essential for acoustic modeling—is not appropriate because of the high out-of-vocabulary (OOV) rate. Furthermore, Arabic ortho-graphic transcriptions are written without diacritic marks. Diacritic marks are es-sential to estimate short vowels, nunation, gemination, and silent letters. State of the art techniques for MSA phonetic transcription are usually done in several phases. In one phase, transcriptions are written without diacritics. Afterwards, automatic di-acritization is performed to estimate missing diacritic marks (WER is 15%–25%) as in (Kirchhoff and Vergyri 2005) and (Sarikaya et al. 2006). Finally, the mapping from diacritized (also called vowelized) text to phonemes is almost a one-to-one mapping.

Dialectal Arabic usually differs significantly from MSA to the extent to con-sider dialects as totally different languages. That is why phonetic transcription tech-niques for MSA cannot be applied directly on dialectal Arabic. In order to avoid automatic or manual diacritization, graphemic acoustic modeling was proposed for MSA (Billa et al. 2002) where the phonetic transcription is approximated to be the sequence of word letters whilst ignoring short vowels. However, the performance is still below the accuracy of phonemic models. Since MSA and the Arabic dialects share the same character inventory, the grapheme-based approach was also applica-ble for dialectal Arabic as shown in (Vergyri et al. 2005) and (Afify et al. 2006).

Exisiting speech corpora for dialectal Arabic are very sparse and suffer from several problems like low recording quality, very spontaneous speech, and they are usually not provided with phonetic transcription. Examples for the existing dialectal corpora are in (Canavan et al. 1997; Makhoul et al. 2005; Maamouri et al. 2006, 2007; Appen 2007, 2006a, 2006b).

In (Canavan et al. 1997), a speech corpus for *Egyptian Colloquial Arabic* (ECA) was developed. The corpus consists of international telephone conversations be-tween native ECA speakers. All calls were originated in North America and were placed to Egypt. Most participants called family members or close friends. The pro-posed corpus in (Canavan et al. 1997) suffers from several problems. Firstly, it was recorded in lowband telephony quality and includes a lot of transmission noise, thus lacks high speech quality that is needed for microphone-based speech applications. Secondly, the speech is very spontaneous and includes a lot of filler noises like hes-itations, laughing, truncated words, coughing, etc.

In (Makhoul et al. 2005), a speech corpus for *Levantine Colloquial Arabic* (LCA) was developed. The corpus consists of microphone-based speech recordings. These were performed in high-band using high quality recording setup. The proposed cor-pus in (Makhoul et al. 2005) was provided without any phonetic transcriptions nor lexicons. Hence, phoneme-based acoustic modeling is not possible.

Some efforts were done to develop ASR systems for dialectal Arabic relying only on the little existing dialectal data as in (Vergyri et al. 2005; Afify et al. 2006). However, the authors did not benefit from the existing large speech corpora for MSA.

In (Vergyri et al. 2005), a speech recognizer was developed for conversational LCA. The used corpus was not phonetically transcribed, i.e. transcriptions do not include diacritic marks. That is why an approximation was performed by ignoring all the phonemes that are estimated from missing diacritic marks and a grapheme-based acoustic model was trained.

In this research, our first goal is to find a way to benefit from existing high quality MSA speech corpora to improve dialectal Arabic speech recognition. The second goal is to find a novel and reliable approach to phonetically transcribe dialectal speech data. It was also necessary to collect a high quality speech corpus for dialectal Arabic. This corpus can be used as a reference to evaluate any proposed approach for imoproving ASR for dialectal Arabic.

Up to our knowledge, there is only one previous work that tackled our first goal. In order to benefit from existing MSA speech resources (Kirchhoff and Vergyri 2005), a cross-lingual approach was proposed where the authors tried to use a data pool of MSA and dialectal speech data in training the acoustic model. MSA training data was a read-speech news broadcast corpus whilst dialectal data was spontaneous telephone calls. The result was a reduction in Word Error Rate (WER) from 42.7% to 41.4% (relative: −3%). We think that the problem with the proposed approach in (Kirchhoff et al. 2002; Kirchhoff and Vergyri 2005) is that the more the data of MSA, the less contribution of ECA. Statistically, a small ECA corpus may not be sufficient to change the acoustic model parameters in conjunction with a huge amount of MSA data. The larger amount of MSA data will have the dominant effect and mask all the acoustic characteristics found in a small dialectal speech amount. Hence, the model will be always biased to MSA and adding more MSA data will decrease the contribution of the little amount of dialectal data. Moreover, we cannot benefit from MSA speech data to train the phonemes in dialectal Arabic that do not exist formally in MSA.

Regarding the problem of phonetic transcription for dialectal Arabic, in (Canavan et al. 1997), a Romanization notation was proposed for Arabic phonetic transcription. However, the proposed technique needs specifically trained transcribers that seems to be too costly given a significant long development time.

1.2 Motivation and Contribution

In order to tackle the problem of the limited availability of dialectal Arabic speech resources, basically, we are proposing a cross-lingual acoustic modeling approach for dialectal Arabic, where we can benefit from existing MSA speech resources. This should improve the dialectal Arabic recognition rate. Our assumption in this work is that MSA is always a second language for any Arabic speaker. In most cases we can identify the original dialect of a speaker even though he is speaking MSA. For instance, in MSA news broadcast, Egyptian speakers usually tend to use the phoneme /g/ instead of /ʤ/ while speaking in MSA and this is a clue that most likely the speaker is originally Egyptian. Consequently, we assume that an acoustic model trained with a sufficient number of MSA speakers will implicitly model the acoustic features of the different Arabic dialects, so we can call it dialect-independent acoustic modeling. In order to fit the MSA acoustic model with a specific Arabic dialect (i.e. to make it dialect-dependent), we can use state-of-the-art acoustic modeling adaptation techniques like Maximum Likelihood Linear Regression (MLLR) (Leggetter and Woodland 1995), Maximum A-Posteriori (MAP) re-estimation (Lee and Gauvain 1993), or a combination of different techniques. The proposed cross-lingual acoustic modeling approach (Elmahdy et al. 2010) showed a significant improvement in speech recognition accuracy compared to previous work in (Rabiner 1989).

In order to tackle the problem of phonetic transcription for dialectal Arabic, we propose a novel approach that benefits from the Arabic Chat Alphabet (ACA) instead of the classical Arabic orthography (Elmahdy et al. 2011). According to our knowledge, there is no previous work that tried to use the ACA for dialectal Arabic speech transcription. Previous work in dialectal Arabic phonetic transcription either focused on implementing their own phonetic transcription conventions and standards as in (Canavan et al. 1997) which is a very costly process. Other researches focused on graphemic transcription as in (Maamouri et al. 2004; Vergyri et al. 2005; Afify et al. 2006) rather than the actual phonetic transcription. The graphemic transcription approach lacks important sounds like short vowels, gemination, etc.

In our research, Egyptian Colloquial Arabic (ECA) and Levantine Colloquial Arabic (LCA) have been chosen as two typical Arabic dialects. Throughout this book, ECA is meant to be as spoken in Cairo.

1.3 Overview

- In Chapter 2, we have included a comprehensive introduction about the Arabic language from an ASR point of view. Furthermore, we have presented the fundamentals of ASR in general.
- In Chapter 3, we discuss how speech datasets have been collected and selected. Speech data sets will be later used to evaluate all proposed approaches.
- In Chapter 4, cross-lingual phonemic acoustic modeling methodologies and results are presented.
- In Chapter 5, cross-lingual graphemic acoustic modeling methodologies and results are presented.
- In Chapter 6, Arabic Chat Alphabet based acoustic modeling is evaluated versus phonemic and graphemic modeling.
- In Chapter 7, all results and contributions are summarized and future work is proposed.

Chapter 2
Fundamentals

2.1 Introduction

Working in the area of Arabic speech recognition requires a good knowledge about the characteristics of the Arabic language and the principals of automatic speech recognition (ASR) systems. In this chapter, the fundamentals of ASR are introduced. Furthermore, we highlight the characteristics of the Arabic language with relevance to ASR. This includes, characteristics on the acoustic and the language level of Arabic that have to be taken carefully into consideration.

2.2 Automatic Speech Recognition

2.2.1 Speech Recognition Architecture

The goal of an ASR system is to map from an acoustic speech signal (input) to a string of words (output). In speech recognition tasks, it is found that spectral domain features are highly uncorrelated compared to time domain features. The Mel-Frequency Cepstral Coefficient (MFCC) is the most widely used spectral representation for feature extraction.

Actually, speech is not a stationary signal, that is why we cannot calculate the MFCCs for the whole input speech signal O. We divide O into small enough overlapping frames o_i where the spectral is always stationary. The frame size is typically 10–25 milliseconds. The frame shift is the length of time between successive frames. It is typically 5–10 milliseconds.

$$O = o_1, o_2, o_3, \ldots, o_t \qquad (2.1)$$

Similarly, the sentence W is treated as if it is composed of a string of words w_i:

$$W = w_1, w_2, w_3, \ldots, w_n \qquad (2.2)$$

M. Elmahdy et al., *Novel Techniques for Dialectal Arabic Speech Recognition*,
DOI 10.1007/978-1-4614-1906-8_2, © Springer Science+Business Media New York 2012

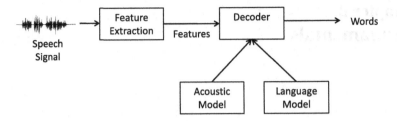

Fig. 2.1 High level block diagram for a state-of-the-art automatic speech recognition system

Now, the task of an ASR system is to estimate the most probable sentence $\hat{W}$ out of all sentences in the language L given the input speech signal O. This can be expressed as:

$$\hat{W} = argmax_{W \in L} P(W|O) \tag{2.3}$$

Since there is no direct solution for the above equation, Bayes rule is applied to break Eq. (2.3) down as:

$$\hat{W} = argmax_{W \in L} \frac{P(O|W)P(W)}{P(O)} \tag{2.4}$$

Actually, the probability of observations $P(O)$ is the same for all possible sentences, that is why it can be safely removed since we are only maximizing all over possible sentences:

$$\hat{W} = argmax_{W \in L} P(O|W)P(W) \tag{2.5}$$

$P(W)$ is the probability of the sentence itself and it is usually computed by the language model. While the conditional probability $P(O|W)$ is the observation likelihood and it is calculated by the acoustic model. See Fig. 2.1 for a high level digram of an ASR system.

2.2.2 Language Modeling

The probability of the sentence (string of words) $P(W)$ is defined as:

$$P(W) = P(w_1, w_2, w_3, \ldots, w_n) \tag{2.6}$$

Usually $P(W)$ is approximated by an N-gram language model as follows:

$$P(w_1^n) \approx \prod_{k=1}^{n} P(w_k | w_{k-N+1}^{k-1}) \tag{2.7}$$

Where n is the number of words in the sentence and N is the order of the N-gram model. To estimate the probability of the different N-grams in the language, a large text corpus is required to train the language model. Typically in ASR, the language model order N is either two (bi-gram) or three (tri-gram).

Fig. 2.2 Left-to-right (Bakis)
hidden Markov Model
(HMM)

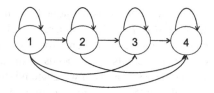

Fig. 2.3 Three states HMM
to model one phoneme, where
a_{ij} is the transition weight
from state i to j

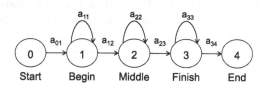

2.2.3 Acoustic Modeling

The conditional probability $P(O|W)$ is computed by the acoustic model. It is typically a statistical model for all the phonemes in the language to be recognized. This type of acoustic models is known as phoneme-based acoustic models. Statistical modeling for the different phonemes is performed using Hidden Markov Model (HMM). Since the speech signal is causal, the left-to-right HMM topology is used for acoustic modeling as shown in Fig. 2.2.

Each phoneme has three sub-phones (begin, middle, and finish), that is why each phoneme is typically modeled using three HMM states besides two non-emitting states (start and end). Skip-state transitions are usually ignored to reduce the total number of parameters in the model as shown in Fig. 2.3.

The same phoneme can have different acoustic characteristics in different words. The phoneme /t/ is pronounced in different ways in the words: tea, tree, city, beaten, and steep. The process by which neighboring sounds influence one another is called *Co-articulation Effect*. Modeling phonemes without taking into consideration the context (left and right phonemes) is called *Context Independent* (CI) acoustic modeling where each phoneme (or sub-phone) is modeled using only one HMM.

Acoustic models for each phoneme usually model the acoustic features of the beginning, middle, and end of the phoneme. Usually the starting state is affected by the left phoneme and the final state is affected by the right phoneme. In *Context Dependent* (CD) acoustic models (tri-phones), co-articulation effect is taken into consideration. For the above /T/ example, five different tri-phones models are built because the left and right context of /T/ is not the same in the five words. Actually, CD acoustic modeling leads to a huge number of HMM states. For example, a language with a phoneme set of 30 phonemes, we would need 27,000 tri-phones. The large number of tri-phones leads to a training data sparsity problem. In order to reduce the total number of states, clustering of similar states is performed and *State Tying* is applied to merge all states in the same cluster into only one state. The merged states are usually called *Tied States*. The optimal number of tied states is usually set empirically and it function in the amount of training data and how much tri-phones are covered in the training data.

The most common technique for computing acoustic likelihoods is the *Gaussian Mixture Model* (GMM). Where each HMM state is associated with a GMM that models the acoustic features for that state. Each single Gaussian is a multivariate Gaussian where its size depends on the number of MFCCs in the feature vector. Basically, GMM consists of weighted mixture of n multivariate Gaussians. For speaker independent (SI) acoustic modeling, the number of multivariate Gaussians n is typically ~ 8 or higher. The more number of Gaussians, the more accents that can be modeled in the acoustic model and the more data that is required to train the acoustic model.

To summarize, an acoustic model is defined by the following for each HMM state:

- Transition weights.
- Means and variances of multivariate Gaussians in the GMM.
- Mixture weights of multivariate Gaussians in the GMM.

2.2.4 Training

All acoustic model parameters that define the acoustic model are estimated using training data. Training data is a collection of large number of speech audio files along with the corresponding phonetic transcriptions. Phonetic transcriptions are usually not written explicitly for each audio segment. However, orthographic transcriptions are written and phonetic transcriptions are estimated using a lookup table.

The mapping from orthographic text to phonemes is usually performed using a lookup table called pronunciation dictionary or lexicon. A pronunciation dictionary includes all the words in the target language and the corresponding phonemes sequence. The word HMM is simply the concatenation of the phoneme HMMs. Since the number of phonemes is always finite (50 phonemes in American English), it is possible to build the word HMM for any given word. In Fig. 2.4, the mapping from orthographic transcription to phonetic transcription is clarified.

In order to train a speaker independent acoustic model, a large number of speakers is required (~ 200) in order to be able to model the variability across speakers.

2.2.5 Evaluation

Word error rate (WER) is a common evaluation metric of the performance of ASR systems and many other natural language processing tasks. The WER is based on how much the word string returned by the recognizer (called the hypothesized word string) differs from a correct or reference transcription. The general difficulty of measuring performance is due to the fact that the recognized word string may have a different length from the reference word string (the correct one). This problem is tackled by first aligning the recognized word sequence with the reference word

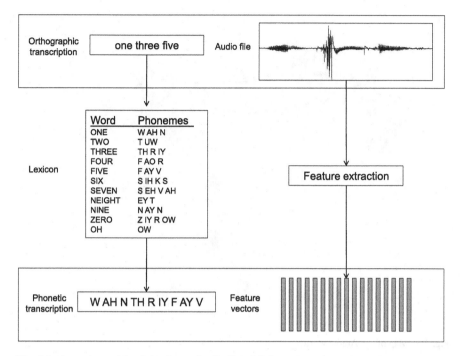

Fig. 2.4 Acoustic model training inputs (audio file of digits and corresponding orthographic transcription). The phonetic transcription is estimated from the lexicon and aligned with the feature vectors

sequence to compute the minimum edit distance (minimum error). This alignment is also known as the maximum substring matching problem, which is easily handled by dynamic programming. The result of this computation will be the minimum number of word substitutions, word insertions, and word deletions necessary to map between the correct and the recognized strings. Word error rate can then be computed as:

$$WER = 100 \times \frac{I + S + D}{N} \tag{2.8}$$

Where I, S, and D are the total number of Insertion, Substitution, and Deletion errors respectively, while N is the number of words in the correct (reference) transcriptions.

More details about the theory of speech recognition can be found in (Huang et al. 2001; Jurafsky and Martin 2009; Rabiner 1989).

2.3 The Arabic Language

The Arabic language is the largest still living Semitic language in terms of number of speakers. Arabic speakers exceed 250 million first language speakers and the number of speakers using Arabic as a second language can reach four times that

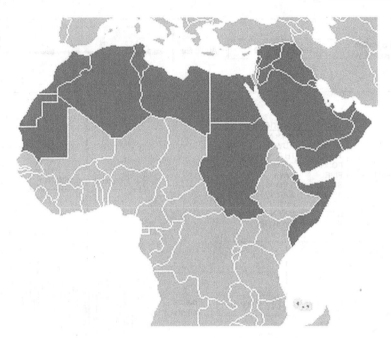

Fig. 2.5 Map of the Arabic world

number. Arabic is the official language in 22 countries known as the Arab world
(see Fig. 2.5) and ranked as the 6th most spoken language based on the number
of first language speakers and it is one of the six official languages of the United
Nations.

The Arabic alphabet is also used in many other languages such as Arabic, Azer-
baijani, Baluchi, Eastern Cham, Comorian, Dogri, Hausa, Kashmiri, Kurdish, Lah-
nda, Pashto, Persian (Iranian and Dari), Punjabi, Sindhi, Uighur, and Urdu.

Arabic script is written from left to write, and some letters may change shape
according to position within the word.

Some work has been done to represent Arabic letters using Roman character sets.
This Roman representation is usually called *transliteration* or *Romanization*. One
of the popular Arabic transliterations is the *Buckwalter* transliteration (Buckwalter
2002a).

The commonly used binary encoding standards for Arabic characters are: ISO
8599-6, the Windows based CP-1256, and the Unicode standard (see Table A.1).

The Arabic alphabet has 28 basic letters (see Table 2.1), not having upper and
lower case like the Roman alphabet. The following is the sentence "The boy went
to the market" in Arabic and the corresponding transliteration:[1]

---

[1]All transliterations in this chapter are following the ZDMG transcription (DIN 31635), generated
by the ArabTEX package unless otherwise mentioned (Lagally 1992).

Table 2.1 The Arabic alphabet

Letter name		Form	Letter name		Form
أَلِف	ʾlif	ا	ضَاض	ḍāḍ	ض
بَاء	bāʾ	ب	طَاء	ṭāʾ	ط
تَاء	tāʾ	ت	ظَاء	ẓāʾ	ظ
ثَاء	ṯāʾ	ث	عَين	ʿayn	ع
جِيم	ǧiym	ج	غَين	ġayn	غ
حَاء	ḥāʾ	ح	فَاء	fāʾ	ف
خَاء	ḫāʾ	خ	قَاف	qāf	ق
دَال	dāl	د	كَاف	kāf	ك
ذَال	ḏāl	ذ	لَام	lām	ل
رَأ	rāʾ	ر	مِيم	mym	م
زَاي	zāy	ز	نُون	nuwn	ن
سِين	syn	س	هَاء	hāʾ	ه
شِين	šyn	ش	وَاو	wāw	و
صَاد	ṣād	ص	يَاء	yāʾ	ي

ḏahaba 'l-walado ʾilaāā 'l-ssuwqi.

ذَهَبَ آلْوَلَدُ إِلَى آلسُّوقِ.

In Arabic there are three diacritic marks that symbolize short vowels; "ـَ" (فَتْحَتْ *fatḥat*), "ـُ" (ضَمَّتْ *dammat*) and "ـِ" (كَسْرَتْ *kasrat*). In addition to these three, there is also the "ـْ" (سُكُون *sukuwn*) which symbolizes the absence of a vowel, the "ـّ" (شَدَّتْ *šaddat*), which symbolizes the duplication of a consonant, and the three "ـً", "ـٌ" and "ـٍ" (تَنوِين *tanwiyn*) marks which have the effect of adding one of the short vowels followed by "n", and can only be placed at the end of the word. These marks are shown in Table 2.2. It should also be noted that the shadda marks are always followed by a *dammat*, *fatḥat*, *kasrat* or a *tanwyn* mark. This means that one letter can be followed by more than one mark.

The Arabic language has many varieties. These are classified into two main classes: Standard Arabic and Dialectal Arabic as shown in Fig. 2.6. Standard Arabic includes *Classical Arabic* and *Modern Standard Arabic (MSA)* while dialectal Arabic includes all forms of spoken Arabic in everyday life communications. Arabic dialects significantly vary among countries and deviate from standard Arabic to some extent and even within the same country we can find different dialects. While there are many forms of Arabic, there are still many common features on the acoustic level and the language level.

In this chapter, we have chosen classical Arabic and MSA forms to highlight the characteristics of standard Arabic. For dialectal Arabic we have chosen Egyptian

Table 2.2 The Arabic
diacritic marks

Diacritic name	Form
فَتَحَت *fatḥat*	˶
ضَّمَّت *ḍammat*	˙
كَسَرَت *kasrat*	˷
شَدَّت *šaddat*	˵
سُكُون *sukuwn*	˚
تَنوين فَتَحَت *tanwyn fatḥat*	˶
تَنوين ضَّمَّت *tanwyn ḍammat*	˷
تَنوين كَسَرَت *tanwyn kasrat*	˵

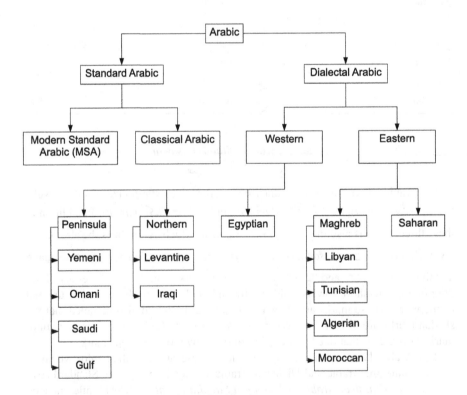

Fig. 2.6 Arabic varieties geographical classification

colloquial Arabic (ECA) (ECA usually stands for the spoken language in Cairo)
as a typical dialectal Arabic form. The main ASR problems for Arabic discussed
in this chapter are: morphological complexity, dialectal forms, and grapheme-to-
phoneme.

2.3.1 Modern Standard Arabic

MSA is also known as Fussha and it is the current standard form of Arabic. Almost all Arabic text resources are written in MSA. It is the formal spoken Arabic language. MSA is used in books, newspapers, news broadcasts, formal speeches, movies subtitling, etc. MSA can be considered as a second language for all Arabic speakers. They can understand spoken MSA as in news broadcasts. MSA is the only commonly accepted Arabic form throughout all native Arabic speakers. That is why many radio and TV broadcasters use MSA in order to target a broad Arabic audience.

The allowed syllables in Arabic are CV, CVC, and CVCC where C is a consonant and V is a long or a short vowel. Therefore, Arabic utterances and words can only start with a consonant and the CVCC pattern is only allowed at the end of a word.

The MSA phonemes inventory consists of 38 phonemes. These include 29 original consonants, three foreign consonants, and six vowels as shown in Tables 2.3 and 2.4. The SAMPA notation is used in our work for the phonetic transcription (Gibbon et al. 1997) in conjuction with the IPA notation. The foreign consonants are /g/, /p/, and /v/. They are rarely found in MSA and appear in loan words. The phoneme /l'/ is also rare as it appears only in the word /?al'l'a:h/ (The God) and its derivatives.

Usually the duration of long vowels is approximately twice the duration of short vowels. Long vowels are not usually treated as two successive short vowels because the duration of long vowels is not exactly twice the duration of short ones (Djoudi et al. 1989).

Arabic is characterized by the existence of pharyngeal and emphatic nature of some consonants (El-Halees 1989; Djoudi et al. 1990). Emphatic consonants in Arabic are /XV, /t'/, /d'/, /D'/, and /s'/. Those types of phonemes exist in Semitic languages (Holes 2004). Emphatic sounds have also significant effect over the whole word containing the emphatic consonant. Some research papers showed a smaller difference between F1 and F2 for all vowels in those words containing emphatic consonant in almost any position (Hassan and Esling 2007). Two diphthongs may also be considered in MSA. These are /ay/ (/a/ followed by /y/) and /aw/ (/a/ followed by /w/) where a vowel is followed by a semi-vowel.

MSA speech corpora are mainly available in the domain of news broadcasts at a relatively low price and those corpora can be used to build speaker independent acoustic models (Yaseen et al. 2006; LDC 2010).

In the MSA transcription, foreign phonemes may not be treated as extra sounds. They can be grouped with the closest phoneme as it is difficult to distinguish between them from the Arabic orthographic transcription. For example /f/ and /v/ may be grouped together and treated as the same phoneme. The same approach may be applied on /b/ and /p/, and also applied on /g/, /Z/, and /dZ/.

The reason for this grouping is mainly because foreign phonemes are rarely used in MSA compared to original sounds. Futhermore, standard Arabic letters do not have any standard letter assigned for foreign sounds. Some efforts were done to differentiate between foreign and original sounds by using non-standard Arabic letters

Table 2.3 Consonants of Modern Standard Arabic in IPA and SAMPA

IPA	SAMPA	Description
b	b	Plosive, voiced bilabial
t	t	Plosive, voiceless dental plain
d	d	Plosive, voiced dental plain
ṭ	t'	Plosive, voiceless dental emphatic
ḍ	d'	Plosive, voiced dental emphatic
k	k	Plosive, voiceless velar
g	g	Plosive, voiced velar
q	q	Plosive, voiceless uvular
ʔ	?	Plosive, voiceless glottal
p	p	Plosive, voiceless bilabial
f	f	Fricative, voiceless labio-dental
v	v	Fricative, voiced labio-dental
θ	T	Fricative, voicelees interdental plain
ð	D	Fricative, voiced interdental plain
ð̣	D'	Fricative, voiced interdental emphatic
s	s	Fricative, voiceless alveolar plain
z	z	Fricative, voiced alveolar plain
ṣ	s'	Fricative, voiceless alveolar emphatic
ʃ	S	Fricative, voiceless postalveolar
ʒ	Z	Fricative, voiced postalveolar
ʤ	dZ	Affricative, voiced postalveolar
x	x	Fricative, voiceless velar
ɣ	G	Fricative, voiced velar
ħ	X\	Fricative, voiceless pharyngeal
ʕ	?\	Fricative, voiced pharyngeal
h	h	Fricative, voiceless glottal
r	r	Trill, alveolar
l	l	Liquid, dental plain
ḷ	l'	Liquid, dental emphatic
w	w	Approximant (semi vowel), bilabial
j	j	Approximant (semi vowel), palatal
m	m	Nasal, bilabial
n	n	Nasal, alveolar

like using the letter پ for the phoneme /p/, the letter ڤ for the phoneme /v/, and the letter چ for the phoneme /g/. However, these conventions are not standard and even many standard Arabic keyboard layouts do not show these letters and also

Table 2.4 Vowels of
Modern Standard Arabic
in IPA and SAMPA

IPA	SAMPA	Description
i	i	Short vowel, close front unrounded
a	a	Short vowel, open front unrounded
u	u	Short vowel, close back rounded
iː	iː	Long vowel, close front unrounded
aː	aː	Long vowel, open front unrounded
uː	uː	Long vowel, close back rounded

standard Arabic character sets as in the ISO 8859-6 do not include these additional letters (ISO 1987).

2.3.2 Classical Arabic

Classical Arabic is the most formal and standard form of Arabic and it is the language of the Quran (the holy book for Muslims). Classical Arabic script as used in Quran represents almost completely the phonetic transcription of the word because the script is fully vowelized and includes diacritic marks that are usually omitted in MSA orthography.

The Arabic language is characterized by the presence of the emphatic consonant /d'/. It is believed that this phoneme as pronounced in classical Arabic is exclusively appearing in Arabic and not in any other language. That is why Arabic is usually defined as *The language of /d'/* (Newman 2002). The Quran script is also characterized by the presence of the *Alif accent* (also called dagger Alif).

Quran phonetics (according to Hafs's narration from Assim) includes the sounds of MSA (except foreign phonemes) plus some extra sounds. The following summarizes the main extra sounds that can appear in the Quran recitation:

- Vowel prolongation with duration of four, five, or six short vowels.
- Necessary prolongation of six short vowels.
- Obligatory prolongation of four or five short vowels.
- Permissible prolongation of two, four, or six short vowels.
- Nasalization (ghunnah) with duration of two short vowels.
- Emphatic pronunciation of the consonant /r/ may happen depending on the context of the consonant /r/.
- Echoing sound in unrest letters (qualquala) for the consonants /q/, /t'/, /b/, /dZ/, and /d/ may happen depending on the context of those consonants.

Using Quran in speech recognition is currently limited to recitation learning applications as in (Abdou et al. 2006; Razak et al. 2008). In these applications, the acoustic model is trained with Quran recitations and the user is asked to utter a specific verse and then the application identifies recitation mistakes.

2.3.3 Dialectal Arabic

Colloquial (or dialectal) Arabic is the natural spoken Arabic in everyday life communications which is not the case for MSA. Colloquial Arabic is not used as a standard form of Arabic in writing. There are many Arabic dialects and almost every country has its own colloquial form. Even within the same country we can find different dialects. Dialectal Arabic may be classified into two groups: Western and Eastern Arabic. Western Arabic can be subclassified into Moroccan, Tunisian, Algerian, and Libyan dialects, while Eastern Arabic can be subdivided into Egyptian, Gulf, Damascus, and Levantine. The Damascus Arabic is considered the closest dialect to MSA. Arabic speakers with different dialects usually use MSA to communicate.

Since dialectal Arabic is not formally written and does not have any orthographic standard, transcribing adequate speech corpora for dialectal Arabic for the purpose of acoustic modeling is very costly.

For large vocabulary tasks, it is also difficult to obtain a large dialectal text corpus (in the order of millions of words) that is necessary in statistical language modeling.

The available corpora for dialectal Arabic are expensive compared to MSA and they are either low quality telephony conversations corpora as the CALLHOME Egyptian Arabic Speech (Canavan et al. 1997), or read speech corpora but with limited vocabulary as in the Orientel project (Zitouni et al. 2002).

Egyptian Colloquial Arabic

ECA is the first ranked Arabic dialect in terms of number of speakers. Moreover, ECA is the most known dialect among Arabic speakers. The main characteristics of ECA phonetics compared to MSA are:

- /t/ and /s/ are used instead of /T/. E.g. /Tala:Tah/ (three) in MSA is transformed to /tala:tah/ in ECA.
- /g/ is used instead of /Z/ and /dZ/. E.g. /Zami:l/ (beautiful) in MSA is transformed to /gami:l/ in ECA.
- /ʔ/ is used instead of /q/. E.g. /qabl/ (before) in MSA is transformed to /ʔabl/ in ECA.
- The existence of the mid front unrounded long vowel /E:/ and short vowel /E/ (they do not exist in MSA). E.g. /ʔiTnayn/ (two) in MSA is transformed to /ʔitnE:n/ in ECA.
- The existence of the open back unrounded long vowel /A:/ and short vowel /A/ (they do not exist in MSA). E.g. /ʔarbaʔ\ah/ (four) in MSA is transformed to /ʔArbAʔ\Ah/ in ECA.
- The existence of the mid back rounded long vowel /O:/ (it does not exist in MSA). E.g. /jawm/ (day) in MSA is transformed to /jO:m/ in ECA (Hinds and Badawi 2009; Stevens and Salib 2005).

On the lexical level, some words exist only in ECA and not in MSA like: /t'ArAbE:zA/ (table) in ECA while it is /t'awila/ in MSA. The sentence structure in ECA tends to be VSO while in MSA it tends to be SVO.

The transcription of ECA is difficult because people are quite influenced by MSA and always write the MSA word instead, for example the ECA word /tamanjah/ (eight) is usually wrongly transcribed as /tama:njah/ or /Tama:njah/ and keep including the long vowel /a:/ as in MSA while it is replaced in ECA by the short vowel /a/.

2.3.4 Morphological Complexity

Arabic is a morphological very rich language and hence dealing with Arabic as morphemes rather than words will limit the size of vocabulary and significantly decrease the number of Out-Of-Vocabulary (OOV) words. For instance a lexicon of 65,000 words in the domain of news broadcast leads to an OOV rate of 4% in Arabic whilst in English it leads to an OOV rate of less than 1%. Moreover, increasing the size of the lexicon substantially might only decrease the OOV by a small percentage. For example, a lexicon of 128,000 words will only lower the OOV rate to 2.4% (Billa et al. 2002).

Most words in Arabic have a root that consists of three consonants called radicals (rarely two and four). A large number of affixes (prefixes, infixes, and suffixes) can be added to the three consonant radicals to form patterns. Arabic is a highly inflected language with gender, number, tense, person, and case. A single Arabic word can represent a whole English sentence like: وَبِاسْتِطَاعَتِهِم /wabi?stit'a:?\atihim/ (and with their ability). Table 2.5 explains the morphological complexity in Arabic by showing some of the affixes that can be added to the word لَاعِب (player), and how this changed the meaning of the word.

Since Arabic is a morphologically rich language, a simple lookup table is not practical for phonetic transcription nor in semantic analysis. The high number of morphemes leads to an exponential increase of the number of words in Arabic. The total number of Arabic words is as large as 6×10^{10} due to the morphological complexity as reported in (Darwish 2002). That is why significant effort has been investigated in the area of morphological analysers for Arabic as in (Buckwalter 2002b; Xiang et al. 2006) and Arabic stemmers to estimate word root as in (Alshalabi 2005).

2.3.5 Diacritization

A strong grapheme-to-phoneme (almost one-to-one mapping) relation only applies for the diacritized Arabic script. Diacritics appear only in sacred books (like Quran) and Arabic language teaching books. MSA is usually written with the absence

Table 2.5 A few affixes of the word لَاعِب *lāᶜib* (player)

Word	Transliteration	Meaning
لَاعِب	*lāᶜib*	player
أَللَّاعِب	*al-lāᶜib*	the player
لَاعِبَان	*lāᶜibān*	two players
لَاعِبِين	*lāᶜibiyn*	players
وَاللَّاعِب	*wa-'l-lāᶜib*	and the player
فَاللَّاعِب	*fa-'l-lāᶜib*	so the player
كَاللَّاعِب	*ka-'l-lāᶜib*	like the player
بِاللَّاعِب	*bi-'l-lāᶜib*	with the player
لِلَّاعِب	*li-llāᶜib*	to the player
لَاعِبِي	*lāᶜibiy*	my player
لَاعِبُكَ	*lāᶜibuka*	your player
لَاعِبُهُ	*lāᶜibuhu*	his player
لَاعِبُهَا	*lāᶜibuhā*	her player

of diacritic marks and the reader infers missing diacritics from the context. Table 2.6 shows a simple lookup table to convert from orthographic Arabic letters (graphemes) to the corresponding phonemes in SAMPA notation. Diacritic marks are shown in Table 2.7 together with the corresponding phonemes in SAMPA.

Non-diacritized Arabic script leads to a high degree of ambiguity in pronunciation and meaning. For instance, the non-diacritized word كتب may have different possible diacritizations and for each case a different pronunciation and a different meaning as shown in Table 2.8. Another problem is that the vowel diacritic case marker on the last letter in a word is determined by the word position in the sentence for example whether the word is subject or object. Furthermore, the speaker is free to choose whether to pronounce or to omit the vowel case marker.

Actually the missing diacritics is not only a problem in Arabic speech recognition. Missing diacritics is more crucial in Arabic speech synthetic Text-To-Speech (TTS) systems Gu et al. (2007). Estimation of the correct diacritization will reduce ambiguity in all Arabic speech and language processing tasks. The majority of Arabic speech corpora is available with non-diacritized transcription and hence acoustic modeling for the speech recognition task is not straight forward. Therefore, an automatic diacritizer is needed to estimate those missing diacritic marks.

Previous work in Arabic acoustic modeling as reported in (Vergyri and Kirchhoff 2004) include an important stage for automatic script diacritization. Automatic diacritization is also known as automatic vocalization. Many research studied how to automatically estimate missing diacritics from the context in order to determine the exact phonetic transcription (Google 2010; Vergyri and Kirchhoff 2004; Sarikaya et al. 2006; Gal 2002; Habash and Rambow 2007; Nelken and Shieber

Table 2.6 Arabic grapheme-to-phoneme conversion

Grapheme	Phoneme	Grapheme	Phoneme
أ	?	س	s
إ	?	ش	S
ؤ	?	ص	s'
ئ	?	ض	d'
ء	?	ط	t'
آ	?a:	ظ	D'
ى	a:	ع	?\
ا (wasla)	?	غ	G
ا	a:	ف	f
ب	b	ق	q
ت	t	ك	k
ث	T	ل	l
ج	Z	م	m
ح	X\	ن	n
خ	x	ه	h
د	d	و	w
ذ	D	ي	j
ر	r	ة	t
ز	z		

2005). However, the problem still remains unsolvable and the WER of automatic diacritization systems ranges between 15% and 25%.

Automatic diacritization is usually based on statistical language modeling to estimate the most probable diacritics for a word given the context in which the word appears. Automatic diacritization is a very important area in Arabic speech recognition since it was proven that a fully vowelized Arabic script improves accuracy over non-vowelized script (grapheme-based) (Afify et al. 2005). The majority of research in automatic diacritization is usually performed for the tasks of TTS or word disambiguation as in (Sarikaya et al. 2006; Gal 2002; Habash and Rambow 2007; Nelken and Shieber 2005). On the other hand, for the task of acoustic modeling, fewer research have been performed as in (Vergyri and Kirchhoff 2004). A high level digram for an automatic diacritizer is shown in Fig. 2.7.

The available commercial applications for the task of automatic diacritization as Fassieh (RDI 2007) still need manual review in order to achieve lower WER. Manual reviewing is mandatory in order to achieve an accuracy closer to ~99%. However, it is a time consuming and a costly operation especially when preparing large diacritized corpora. The productivity of a well trained linguist is ~1.5K words per work-day (Atiyya et al. 2005). In other words, for a 40 hours speech corpus that consists of 200K words in average, we will need 133 work-days for reviewing the results of a commercial class automatic diacritization system.

Table 2.7 Arabic diacritics to phoneme conversion, example with the consonant /b/

Diacritic	Phoneme	Diacritic description
بَ	b a	Fatha (short vowel a)
بِ	b i	Kasra (short vowel i)
بُ	b u	Damma (short vowel u)
بْ	b	Sukun (no vowel)
بَّ	b b a	Shadda (double consonant) & Fatha
بِّ	b b i	Shadda & Kasra
بُّ	b b u	Shadda & Damma
بًا	b a n	Tanween (Nunation)[2] Fatha
بٍ	b i n	Tanween Kasra
بٌ	b u n	Tanween Damma
بًّا	b b a n	Shadda, Tanween Fatha
بٍّ	b b i n	Shadda, Tanween Kasra
بٌّ	b b u n	Shadda, Tanween Damma
بَا	b a:	Alif Al-Madd (long vowel a:)
بِي	b i:	Yaa Al-Madd (long vowel i:)
بُو	b u:	Waaw Al-Madd (long vowel u:)

Table 2.8 Possible diacritizations for the word كتب with English translation

Word	SAMPA	English meaning
كَتَبَ	/kataba/	He wrote
كُتِبَ	/kutiba/	It was written
كَتَّبَ	/kattaba/	He dictated
كُتِّبَ	/kuttiba/	It was dictated
كُتُب	/kutub/	Books

2.3.6 Challenges for Dialectal Arabic Speech Recognition

Actually, training an acoustic model for dialectal Arabic is very challenging. In order to train an acoustic model for a particular Arabic dialect, a large speech data set should be collected for that dialect. Unfortunately, dialectal speech data collection is too difficult compared to MSA and compared to other languages like English. The difficulties are mainly due to the inability to estimate the accurate phonetic transcription for dialectal Arabic. This can be clarified in the following:

- A pronunciation dictionary is required to map words to corresponding phoneme sequence. However, due to the high degree of morphological complexity in Ara-

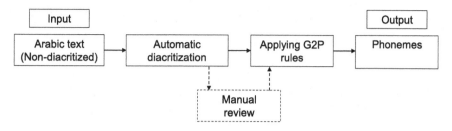

Fig. 2.7 High level block diagram for an Arabic automatic diacritization system

bic, it is impossible to create a pronunciation dictionary that contains all Arabic
words.

- Dialectal Arabic is only spoken and not formally written. There is no commonly
  accepted standard for dialectal Arabic orthography. For instance, the same pronounced letter can be transcribed differently across different transcribers. In other
  words, from the orthographic transcription, it is not possible to estimate the correct pronunciation.
- Since dialectal Arabic is written without diacritic marks, it is not possible to estimate missing short vowels. Furthermore, automatic diacritization systems do not
  exist for dialectal Arabic.

2.4 Summary

In this chapter, an overview about the Arabic language was provided from a speech
recognition point of view. First, we introduced a brief overview about automatic
speech recognition. Second, we provided an overiew about the Arabic language. We
have seen that there exist a large number of Arabic varieties. These were classified into standard and dialectal. Dialectal Arabic represents all the different Arabic
forms as spoken in all Arabic countries. Dialectal Arabic is only spoken in everyday life and there is no standard to represent it in written form. Modern Standard
Arabic (MSA) is currently the standard form of Arabic used in all formal situations
and it is also a second language for all Arabic speakers. Egyptian Colloquial Arabic (ECA) was chosen as a typical dialectal form and we have shown that there
are significant differences between ECA and MSA. Arabic is a morphological very
rich language and a simple lookup table for phonetic transcription is not practical
because of the high OOV rate. Grapheme-to-phoneme (G2P) is difficult for Arabic
because of missing diacritic marks that are not usually written. Therefore, automatic
or manual diacritization is needed in order to estimate the accurate phonetic transcription.

Chapter 3
Speech Corpora

3.1 Introduction

In this chapter, speech corpora that are used in our work are presented and compared including recording settings and transcription conventions. Basically, we made use of three speech corpora, one corpus for Modern Standard Arabic (MSA) and two corpora for dialectal Arabic. The two Arabic dialects that have been chosen in this research are Egyptian Colloquial Arabic (ECA) and Levantine Colloquial Arabic (LCA).

3.2 Modern Standard Arabic Corpus

The majority of MSA speech corpora are commercially available with non-diacritized transcriptions as in (Paulsson et al. 2009; Maamouri et al. 2006). This means that phonetic transcriptions are not directly estimated from orthographic transcriptions.

For non-diacritized MSA transcriptions, a morphological analyzer such as the *Buckwalter Morphological Analyzer* (Buckwalter 2002b) generates all possible diacritization forms for a given word. This results in multiple pronunciation variants. Afterwards, force-alignment is performed to choose the correct pronunciation to train the acoustic model. The problem with this approach is that Arabic has a high homograph rate. Hence, the same word has more than one possible pronunciation variant reducing the distance between the different pronunciations. This results in more alignment errors and hence reduces the recognition accuracy.

In order to avoid manual or automatic diacritization (since automatic diacritization is not the main focus of our work) as previously described in Section 2.3.5, our aim was to search for an existing MSA corpus that is already supplied with accurate and manually reviewed diacritization.

M. Elmahdy et al., *Novel Techniques for Dialectal Arabic Speech Recognition*,
DOI 10.1007/978-1-4614-1906-8_3, © Springer Science+Business Media New York 2012

3.2.1 Corpus Specifications

The Nemlar news broadcast speech corpus was chosen as a typical MSA corpus (Yaseen et al. 2006; Nemlar 2005; Maegaard et al. 2004). The corpus is provided by the European Language Resources Association (ELRA 2010). It consists of 40 hours of MSA news broadcast speech. The broadcasts were recorded from four different radio stations as follows:

- Radio Orient—94.3 MHz in Paris.
- RMC (Radio Monte Carlo)—88.6 MHz in Paris.
- Medi1—97.60 MHz in Rabat.
- RTM (Radio Television Maroc)—91.10 MHz in Rabat.

Each broadcast contains between 25 and 30 minutes of news and interviews. The recordings were carried out at three different periods between 30 June 2002 and 18 July 2005. All files were recorded in linear PCM format, 16 kHz, and 16 bit. The total number of speakers is 259 and the lexicon size is 62K distinct words with a phoneme set of 34 phonemes. The lexicon is provided with transliterations, word frequency and SAMPA for Arabic is also included.

The radio receiver used for this project was Sangean ATS 909 connected to line in of a Soundblaster compatible sound card. The software Cybercorder 2000 from Skyhawk Technologies was used for scheduling and managing the recordings. The software used for the transcription is Transcriber (Barras et al. 2000) with the additional patch for Arabic. Thus the transcriptions were done in Arabic characters and the software automatically generated the transliterations. The following annotation levels are included:

- Orthographic transcription of speech (in news, not in music, commercials, etc.), including Named Entities.
- Speakers and speaker turns.
- Segment markers (portions of maximum 10 seconds).
- Topic/story boundaries.
- Background noises (stationary and instantaneous noise events).
- Change of background.
- Music/Noise.
- Word boundaries.

This corpus was mainly selected because of several reasons:

- It is a relatively new MSA corpus.
- The transcription is fully vowelized—i.e. contains diacritic symbols—and hence we have accurate phonetic transcription. Thus, there is no need to use automatic or manual diacritizer as in (Vergyri and Kirchhoff 2004).
- The high number of speakers will help to better modeling different Arabic accents.
- The corpus is available for a comparatively low price.
- Different types of noise are accurately labeled. Hence, it is easy to postprocess the corpus to exclude low quality or non-speech segments.

Table 3.1 Specifications summary of the Egyptian colloquial Arabic speech corpus

Arabic variety	standard	No. of utterances	41816
Speech type	newsbroadcast	Lexicon size	62K
Sampling	16 kHz	Diacritization	yes
Quantization	16 bit	Phonemic lexicon	yes
No. of speakers	259	DTC	57K
Total time (hrs)	33	Phoneme set size	34

3.2.2 Post-Processing

News broadcast corpora contain many types of noise like background music, paper noise, etc. This is due to the fact that they are not originally created for the task of speech recognition. We have post-processed the Nemlar corpus to exclude speech segments with the following types of noise: music, cross-talks, telephone calls, and undetermined noise. We have also excluded speech segments for non-native speakers and those containing truncated words from the corpus. We have kept inspiration, microphone, rustling of paper, and hesitation noise to be included in the filler dictionary since theses effects already exist in the orthographic transcription. After post-processing the 40 hours of speech have been reduced to 33 hours.

3.2.3 Corpus Key Facts

In order to evaluate the acoustic feature coverage, we are proposing the *Distinct tri-phones coverage (DTC)* metric, where we calculate the total number of distinct tri-phones in the corpus. For the MSA corpus, DTC was found to be 57K distinct tri-phones. A summary of the corpus key facts is provided in Table 3.1.

3.3 Egyptian Colloquial Arabic Corpus

The current available speech resources for dialectal Arabic are very finite. For instance, for Egyptian Colloquial Arabic (ECA), there are only few commercial corpora available like (Canavan et al. 1997) and (Zitouni et al. 2002). They are all recorded over telephone lines in 8 kHz, and hence we still lack high quality datasets. That is why we decided to collect our own ECA corpus. In this work, ECA is meant as Arabic spoken in Cairo.

ECA has been chosen as a typical Arabic dialect as it is the most famous and well-known Arabic dialect especially because of the popularity of Egyptian drama in the Arab world. Moreover, there exist some previous work in pronunciation dictionaries as in (Hinds and Badawi 2009) and (Stevens and Salib 2005) that can be helpful in training phonemic acoustic models.

Arabic:	يوم الخميس ١ أكتوبر ٢٠١٠
English:	Thursday 1st of October 2010
IPA:	/yoːm ʔilxamiːs waːhid ʔuktoːbɑr ʔalfeːn wʕɑʃɑːrɑ/
Arabic:	تذكرة درجة أُولَى
English:	First class ticket
IPA:	/tɑzkɑrɑ dɑrɑɡɑ ʔuːla/
Arabic:	بيض مسلوق
English:	Boiled eggs
IPA:	/beːɖ masluːʔ/

Fig. 3.1 Samples from the ECA speech corpus with Arabic, English and IPA transcriptions

3.3.1 Corpus Specifications

A database of the most frequently used ECA words and utterances was created. This database includes utterances from different speech domains like: digits, numbers, greetings, times and dates, word spellings, restaurants, train reservations, Egyptian proverbs, etc. The diversity of speech domains ensures a good coverage of acoustic features. Some samples are shown in Fig. 3.1. Every speaker was prompted 50 utterances randomly generated or selected from the database. Speakers with major lisps have been excluded from the experiment. Speakers were asked to speak naturally with their Egyptian native dialect. The 50 utterances per speaker are categorized as follows:

- Four utterances with a single digit (randomly generated).
- Four utterances with four connected digits (randomly generated).
- Four utterances with four natural numbers (randomly generated).
- Four utterances with four Arabic letters (randomly generated).
- Two utterances with full dates including day of the week (randomly generated).
- Two utterances with time (randomly generated).
- Five utterances with all possible yes/no in ECA.
- One utterance with a rare/foreign ECA phoneme.
- 24 utterances randomly selected from the database (greetings, proverbs, restaurants, reservation, etc.).

3.3.2 Recording Setup

Recordings were performed in a quiet room. Every speaker was given a sheet with all the 50 utterances written in Arabic letters. The speaker was asked to read these utterances while the recording is done continuously in the background. The speaker was asked to repeat any utterance in case hesitations or mistakes are observed.

Table 3.2 Samples from the ECA lexicon with pronunciation variants in IPA

Arabic	English	Pronunciations
ترابيزة	Table	ʈ a r ɑ b eː z ɑ
		ʈ a r ɑ b eː z i t
جراج	Garage	g ɑ r ɑː j
أحمر	Red	ʔa ħm ɑ r
صوت	Sound	ş oː t

The recording system consists of the Sennheiser ME 3-N super-cardioid microphone which is a close proximity microphone with noise cancellation (see Table C.1 for detailed specifications). The polar diagram and the frequency response curve for the ME 3-N microphone are shown in Fig. C.1 and Fig. C.2 respectively. In order to avoid plosive and breathing noises, the microphone was placed about 2.5–3 cm away from the mouth corner. The microphone is connected with the InSync Buddy USB 6G digitizer which is an external USB sound card. The digitizer is based on the Micronas UAC3556b microchip (see Table D.1 for detailed specifications). All recordings were performed in linear PCM, 16 kHz, and 16 bits. To avoid power lines interference a laptop was used. During the recordings, the laptop has been unplugged from the charger and running only on the internal batteries power. The Audacity software was used to record the whole speaker session in one large file. Later on, the file was manually segmented and aligned with the corresponding transcriptions.

3.3.3 Phonetic Transcription

Some phonemes exist in dialectal Arabic but not in MSA. It is therefore not possible to estimate the accurate phonetic transcription from the diacritized text. Moreover, orthographic transcriptions tend to use the original MSA spelling rather than the actual dialectal spelling. That is why we decided to manually create the pronunciation lexicon for ECA.

The lexicon consists of ~700 unique words. It was manually prepared using the dictionaries: *A Dictionary of Egyptian Arabic* (Hinds and Badawi 2009) and *A Pocket Dictionary of the Spoken Arabic of Cairo* (Stevens and Salib 2005). Actually, both dictionaries were published to mainly help tourists and foreigners to learn ECA pronunciation. This is due the popularity of Egypt as a touristic country. Unlike other Arabic dialects, where we could not find any high quality lexicon that we can rely on. The whole developed ECA lexicon is included in Table B.1.

The accurate phonetic transcription was done by aligning the orthographic transcriptions with the corresponding words in the lexicon. Words with more than one pronunciation variants were forced aligned using a context independent acoustic model that was trained using the first variant. Afterwardss the forced aligned transcriptions were manually reviewed in order to correct alignment errors. Pronunciation variants for words with the letters Teh Marbuta and Alef Wasla were added to the lexicon. Samples from the lexicon are shown in Table 3.2. In order to evaluate

Table 3.3 Specifications summary of the Egyptian colloquial Arabic speech corpus

Variety	Egyptian	No. of utterances	1100
Speech type	read	Lexicon size	700
Sampling	16 kHz	Diacritization	no
Quantization	16 bit	Phonemic lexicon	yes
No. of speakers	22	DTC	15K
Total time (hrs)	0.5	Phoneme set size	41

the acoustic features coverage, DTC was found to be 15K distinct tri-phones for our ECA speech corpus.

The phoneme set for ECA consists of overall 41 phonemes, of which 29 are consonants and 12 are vowels. The consonants are: ?, b, p, t, g, ʒ, ħ, x, d, r, z, s, ʃ, ş, ḍ, ţ, ẓ, ʕ, ɣ, f, v, q, k, l, m, n, h, w, and j. The vowels are: a, aː, ɑ, ɑː, i, iː, e, ɑ, u, uː, o, oː.

3.3.4 Corpus Key Facts

Overall we have recruited 22 native Egyptian speakers, 50% of the speakers are males and 50% are females. The speakers are between 18 and 32 years old. The corpus was divided into training/adaptation set of 14 speakers and a testing set of eight speakers. A summary of the corpus key facts have been summarized in Table 3.3.

3.4 Levantine Colloquial Arabic Corpus

In order to ensure the consistency of any proposed approach that improves dialectal Arabic speech recognition, it was necessary to work with more than one dialect. We were able to find one reasonable corpus which is the *BBN/AUB DARPA Babylon Levantine Arabic Corpus* (Makhoul et al. 2005).

3.4.1 Corpus Specifications

The corpus is for spontaneous speech recorded from subjects having Levantine colloquial Arabic as their native language. Levantine Arabic is the dialect of Arabic spoken by people in Lebanon, Syria, Jordan, and Palestine.

This corpus was developed with funding from the Defense Advanced Research Project Agency (DARPA), as part of the Babylon program. The Babylon program is intended to advance the state-of-the-art in speech-to-speech translation systems, both by creating new technology and by developing systems for field use.

Table 3.4 Specifications summary of the Levantine colloquial Arabic speech corpus. Since there is no phonetic transcription with this corpus, it is not possible to calculate the tri-phone coverage

Variety	Levantine	No. of utterances	32864
Speech type	spontaneous	Lexicon size	14.7K
Sampling	16 kHz	Diacritization	no
Quantization	16 bit	Phonemic lexicon	no
No. of speakers	164	DTC	N/A
Total time (hrs)	26	Phoneme set size	N/A

The corpus was recorded using a close-talking, noise-cancelling, headset microphone (the Andrea Electronics NC-65). A data-collection tool, developed by BBN, was used to do the collection. This tool allows the user to select a particular scenario, and then step through the questions. To ask a question, the user clicks a button, and the tool plays out a prerecorded Arabic prompt, corresponding to the Arabic translation of the question. Upon completion of the prompt, the tool goes into the listening mode. The subject utters his reply, which the tool records. When the recording of the subject is finished, the user clicks another button to stop, and then goes on to the next question. Hence, end-pointing of the speech is done manually rather than automatically.

The subjects in the corpus were responding to refugee/medical questions like: (Where is your pain?, How old are you?, etc.), and were playing the part of refugees. Each subject was given a part to play, that prescribed what information they were to give in response to the questions. However, they were advised to express themselves naturally, in their own way, in Arabic. To avoid priming subjects to give their answer with a particular Arabic wording, the parts were given in English rather than Arabic.

The total number of recorded speakers is 164 with a vocabulary size of 14.7K unique words. The lexicon consists of only words in the graphemic form without phonetic transcription. Furthermore, we could not find any accurate pronunciation dictionary to map words to corresponding phonemes. Actually, this case is noraml for all the Arabic dialects since they are only spoken and not written. That is why it is difficult to find a standard way to estimate the correct phonemes sequence for a given dialectal word.

3.4.2 Post-Processing

Since the LCA is very spontaneous, it could be noticed that a large percentage of the corpus contains truncated words and hesitations. On the other hand, both MSA and ECA are not spontaneous. In order to reduce the mismatch between spontaneous and read speech, we have excluded speech segments with labeled filler noises like: coughing, laughing, and hesitations (AA, EH, UM, EM). Segments with truncated words were also excluded from the corpus. Finally after post-processing, the corpus size was reduced from 60 hours to 26 hours. The relevant specifications of the corpus have been summarized in Table 3.4.

3.5 Summary

In this chapter, we have described all the Arabic speech resources used in our research. Since there exist different speech corpora for Modern Standard Arabic (MSA), we have decided to work with one of the existing corpora. Our search criterion was to have a corpus that is distributed already with diacritized Arabic transcription in order to facilitate grapheme-to-phoneme (G2P) conversion.

For dialectal Arabic resources, we could not find high quality corpora that come with an accurate phonetic transcription. That is why we have decided to collect our own Egyptian Colloquial Arabic speech corpus. A high quality recording step was established and all recordings were phonetically transcribed and reviewed. Another corpus for Levantine Colloquial Arabic (LDC) was chosen from the Linguistic Data Consortium (LDC), however the LCA corpus is not phonetically transcribed.

Chapter 4
Phonemic Acoustic Modeling

4.1 Introduction

As previously discussed, one of the major problems in ASR for dialectal Arabic is the sparse speech resources and the limited research done in phonetic transcription. In this chapter, we address the problem of having only a small amount of dialectal Arabic speech data, for which it is possible to have a phonetic transcription.

Egyptian Colloquial Arabic (ECA) has been chosen as a typical dialectal form. In order to improve phonemic acoustic modeling for ECA, we will describe how existing MSA speech resources can be applied to dialectal Arabic speech recognition.

In order to benefit from existing MSA speech resources, a cross-lingual acoustic modeling approach is proposed. It is based on state-of-the-art acoustic model adaptation techniques including MLLR and MAP with our in-house collected ECA corpus. Since MSA and ECA do not share the same phoneme set, it is not possible to apply acoustic model adaptation techniques across MSA and ECA.

In order to make phoneme-based adaptation feasible, we show how to normalize the phoneme sets of MSA and ECA. Speech recognition accuracy using the adapted phonemic acoustic model is compared against a baseline acoustic model that was trained using only a small dialectal corpus. The results are also compared against the data pooling acoustic modeling approach proposed in (Kirchhoff and Vergyri 2005).

4.2 Phonemic Baseline

4.2.1 Baseline System Description

Our speech recognition system is based on the state-of-the-art GMM-HMM architecture (Rabiner and Juang 1993; Rabiner 1989; Huang et al. 2001; Jurafsky and Martin 2009; Young et al. 1996). All acoustic models were trained using the *CMU SphinxTrain* trainer (CMU 2010a) while language models were trained

M. Elmahdy et al., *Novel Techniques for Dialectal Arabic Speech Recognition*,
DOI 10.1007/978-1-4614-1906-8_4, © Springer Science+Business Media New York 2012

using the *CMU-Cambridge statistical language modeling toolkit* (CMU 2010b; Clarkson and Rosenfeld 1997). The *CMU Sphinx4* decoder (Lamere et al. 2003) was used in decoding and evaluating speech recognition accuracy of the different setups.

Different acoustic models were trained using only the ECA training set that consists of 14 different speakers. In order to optimize the number of Gaussian densities and the total number of HMM states, or in other words, the number of tied-states, we have trained several acoustic models with different configurations. No approximations were applied on the phoneme set that consists of 41 phonemes.

For the feature extraction, the feature vectors consist of 39 MFCC features as follows: 12 cepstral coefficients, 12 delta cepstral coefficients, 12 double delta cepstral coefficients, one energy coefficient, one delta energy coefficient, and one double delta energy coefficient. All acoustic models are trained to be all fully continuous density context-dependent tri-phones with three states per HMM.

A statistical bi-gram language model with Kneser-Ney smoothing (Ney et al. 1994) was trained using the orthographic transcriptions of the ECA training set. In order to evaluate the language model, a perplexity test was performed using the ECA testing set transcriptions. Language model testing results were as follows: perplexity of 19.6 (i.e. entropy of 4.29 bits), OOV rate of 6.54%, and bi-gram hits of 81.78%.

In order to decrease the OOV rate, we have prepared additional 25000 text utterances to train the bi-gram model. These utterances have been chosen to be from the same speech domains as in the ECA speech corpus. All language modeling parameters were fixed during the entire experiment, so that any change in speech recognition rate is mainly due to the acoustic model.

4.2.2 Baseline Results

The optimized numbers of tied-states and Gaussians per state were found to be 250 and 4 respectively. Search was done in the range from 62 to 1000 tied-states and from 1 to 32 Gaussians as shown in Table 4.1. For each combination of tied-states and Gaussian densities, a batch recognition test was performed using the ECA testing set that consists of eight different speakers. The speech recognition accuracy has been evaluated in terms of the Word Error Rate (WER) metric after aligning the recognized and the reference text utterances using the minimum edit distance:

$$WER = 100 \times \frac{I + S + D}{N} \tag{4.1}$$

Where I, S, and D are the total number of Insertion, Substitution, and Deletion errors respectively, while N is the number of words in the correct (reference) transcriptions. The result of decoding the ECA test set using the baseline acoustic model was found to be an absolute WER of 13.4% as shown in Table 4.5.

Table 4.1 Word Error Rate (WER) % while optimizing the total number of tied-states (TS) and the number of Gaussian densities for the phoneme-based ECA baseline acoustic models

Gaussians	Tied-states (TS)				
	62	125	250	500	1000
1	31.5%	32.8%	20.0%	16.5%	25.5%
2	23.7%	23.7%	16.2%	15.8%	25.2%
4	19.8%	18.7%	**13.4%**	20.7%	30.5%
8	14.1%	14.5%	13.8%	20.9%	39.6%
16	14.8%	15.4%	17.3%	38.9%	66.3%
32	25.0%	23.0%	30.2%	74.3%	91.3%

4.3   MSA as a Multilingual Acoustic Model

Multilingual ASR represents a demanding research area especially for under-resourced languages as shown in (Fegen et al. 2003; Fung and Schultz 2008; Gruhn and Nakamura 2001; Schultz and Waibel 2001). In this section, our goal is basically to use acoustic models trained with MSA speech data as multilingual models to decode dialectal Arabic speech. For ECA, that was initially impossible because of the mismatch between the phoneme set of MSA and ECA as shown in Fig. 4.1. Despite the existence of a common subset, the problem lies in the phonemes that do exist in ECA and not in MSA. That is why it was not possible to build a HMM model for any word that may include at least one of those phonemes.

Our proposed solution for this problem was to normalize the phoneme sets of both MSA and ECA in such a way to include the whole ECA phoneme set inside the MSA one as shown in Fig. 4.1. Before going into the details of phoneme sets normalization, we will first highlight the major differences between MSA and ECA.

4.3.1   MSA and ECA Phonetic Comparison

Consonants in MSA and ECA are compared in Table 4.2 where we can also find the corresponding Arabic letter for each consonant. Vowels in MSA and ECA are compared in Table 4.3. Short vowels are represented in Arabic text with diacritic marks. In fact, it is not common to write diacritic marks in Arabic orthography as the reader usually infers them from the context. However, in MSA, diacritic marks (if written) uniquely determine a specific vowel. On the other hand, in ECA, a diacritic mark (if written) is not sufficient to uniquely identify the realized vowel as shown in Table 4.3. For instance, the diacritic *Kasra* in ECA may represent an /i/ or an /e/.

We have summarized the main phonetic differences between MSA and ECA as follows:

- There are three short vowels in MSA /a/, /i/, and /u/ besides their long forms /aː/, /iː/, and /uː/ respectively. The /a/ vowel is sometimes pronounced as /ɑ/ (emphatic /a/) but usually it is miss-considered in the phonetic transcription of MSA.

Fig. 4.1 Phoneme sets normalization for MSA and ECA

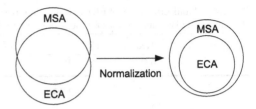

- In ECA, there are three more vowels /e/, /o/, and /ɑ/ along with the corresponding long forms as well.
- The letter ج Jeem should be formally pronounced /ʤ/ in MSA while it is inverted to /g/ in ECA.
- The letter ث Theh is pronounced /θ/ in MSA but it is inverted in ECA to /s/ or /t/.
- The letter ق Qaf is always pronounced /q/ in MSA but in ECA it is sometimes inverted to the glottal stop /ʔ/.
- The letter ذ Thal /ð/ in MSA is inverted in ECA to /d/ or /z/.
- The letter ظ Zah /ð̣/ is pronounced in ECA without being interdental, it is just a /ẓ/ (emphatic /z/).
- There are two diphthongs in MSA /ay/ and /aw/ that do not exist in ECA. They are usually realized in ECA to /eː/ and /oː/ respectively.
- ECA text transcriptions tend to use MSA spelling instead of the exact pronunciation. For example, the word دقيقة (minute) is always written with the letter Qaf /q/ though it is pronounced as a glottal stop /ʔ/.

Ambiguity in Dialectal Arabic Transcriptions   ECA phonetic transcription is usually harder than MSA because the pronunciation of some letters in ECA is phonetically ambiguous. For example, the letter ق Qaf is sometimes pronounced /q/ as in the word قرية (village) and sometimes it is pronounced /ʔ/ in the word أزرق (blue). In MSA, on the other hand, the letter ق Qaf is always pronounced /q/. Other ambiguous letters in ECA are ث Theh and ذ Thal. Table 4.2 shows the different Arabic letters that may represent the same phoneme in ECA.

4.3.2 Phoneme Sets Normalization

Using MSA acoustic models as multilingual models to decode ECA is only possible if the target phoneme set is exactly the same as the MSA set or at least a subset of the MSA phoneme set. The normalization is performed in order to have the same phoneme set and the same phonetic transcription convention across MSA and ECA.

Firstly, we have started to apply several approximations and phoneme mapping rules on the ECA side. The main guideline was to map phonemes to their corresponding origins in MSA as much as possible even if they are totally different (acoustic-wise) as shown in Fig. 4.2. Changes on the ECA phoneme set side were as follows:

Table 4.2 Consonants in MSA and ECA and corresponding Arabic letters

Consonant	Description	MSA Exist	MSA Letters	ECA Exist	ECA Letters
b	Plosive, voiced bilabial	✓	ب	✓	ب
t	Plosive, voiceless dental plain	✓	ت	✓	ت ث
d	Plosive, voiced dental plain	✓	د	✓	د ذ
ṭ	Plosive, voiceless dental emphatic	✓	ط	✓	ط
ḍ	Plosive, voiced dental emphatic	✓	ض	✓	ض ظ
k	Plosive, voiceless velar	✓	ك	✓	ك
g	Plosive, voiced velar	rare	ج	✓	ج
q	Plosive, voiceless uvular	✓	ق	✓	ق
ʔ	Plosive, voiceless glottal	✓	ء	✓	ء ق
p	Plosive, voiceless bilabial	rare	ب	rare	ب
f	Fricative, voiceless labio-dental	✓	ف	✓	ف
v	Fricative, voiced labio-dental	rare	ف	rare	ف
θ	Fricative, voicelees interdental plain	✓	ث	–	–
ð	Fricative, voiced interdental plain	✓	ذ	–	–
ð̣	Fricative, voiced interdental emphatic	✓	ظ	–	–
ẓ	Fricative, voiced alveolar emphatic	–	–	✓	ظ
s	Fricative, voiceless alveolar plain	✓	س	✓	س ث
z	Fricative, voiced alveolar plain	✓	ز	✓	ز ذ
ṣ	Fricative, voiceless alveolar emphatic	✓	ص	✓	ص
ʃ	Fricative, voiceless postalveolar	✓	ش	✓	ش
ʒ	Fricative, voiced postalveolar	rare	ج	rare	ج
ʤ	Affricative, voiced postalveolar	✓	ج	–	–
x	Fricative, voiceless velar	✓	خ	✓	خ
ɣ	Fricative, voiced velar	✓	غ	✓	غ
ħ	Fricative, voiceless pharyngeal	✓	ح	✓	ح
ʕ	Fricative, voiced pharyngeal	✓	ع	✓	ع
h	Fricative, voiceless glottal	✓	ه	✓	ه
r	Trill, alveolar	✓	ر	✓	ر
l	Liquid, dental plain	✓	ل	✓	ل
w	Approximant (semi vowel), bilabial	✓	و	✓	و
j	Approximant (semi vowel), palatal	✓	ي	✓	ي
m	Nasal, bilabial	✓	م	✓	م
n	Nasal, alveolar	✓	ن	✓	ن

Table 4.3 Vowels in MSA and ECA and corresponding diacritic marks

Vowel	Description	MSA		ECA	
		Exist	Diacritic	Exist	Diacritic
a	Low/open and front Fatha muraqqaqa	✓	َ	✓	َ
ɑ	Low/open and back Fatha mufakhkhama	✓	َ	✓	َ
i	High/close and front Kasra khalisa	✓	ِ	✓	ِ
e	Middle/half close and front Kasra mumala	–	–	✓	ِ
u	High/close and back Damma khalisa	✓	ُ	✓	ُ
o	Middle/half close and back Damma mumala	–	–	✓	ُ

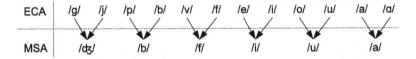

Fig. 4.2 Normalization performed on the ECA phoneme set

Fig. 4.3 Normalization performed on the MSA phoneme set

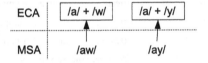

- The vowels /e/, /o/, and /ɑ/ in ECA were approximated to /i/, /u/, and /a/ respectively.
- It was noticed that usually there are errors with the transcription of foreign phonemes in MSA as well as in ECA, for example, the word باريس (Paris) has the foreign phoneme /p/. However it is sometimes miss-interpreted as /b/. That is why we decided to normalize foreign phonemes. Foreign phonemes were normalized to the nearest standard ones, so /v/, /p/, and /ʒ/ were approximated to /f/, /b/, and /ʤ/ respectively.
- The consonant /g/ in ECA was approximated to /ʤ/.
- The consonant /ʐ/ in ECA was approximated to /ð̣/.

Secondly, we tried to benefit from existing MSA phonemes that do not exist in ECA. We found out that the diphthongs in MSA, if treated as two consecutive phonemes as shown in Fig. 4.3, will improve acoustic modeling quality. Changes on the MSA phoneme set side were as follows:

- The diphthongs /aw/ was decomposed and treated as /a/ followed by /w/.
- The diphthongs /ay/ was decomposed and treated as /a/ followed by /y/.

Table 4.4 Word Error Rate (WER) % while optimizing the total number of tied-states (TS) and the number of Gaussian densities using phoneme-based MSA acoustic models

Gaussians	Tied-states (TS)						
	62	125	250	500	1000	2000	4000
1	53.5%	43.5%	39.8%	40.4%	34.3%	37.8%	38.9%
2	42.2%	38.7%	34.8%	34.8%	33.5%	33.7%	34.3%
4	43.0%	38.3%	31.5%	32.2%	28.9%	31.3%	31.1%
8	41.5%	36.1%	29.8%	31.5%	28.5%	28.7%	26.5%
16	37.8%	38.0%	30.7%	27.5%	27.8%	28.9%	29.6%
32	31.1%	32.4%	33.0%	27.8%	**24.5%**	30.0%	32.6%
64	32.2%	32.2%	32.4%	27.4%	28.7%	30.9%	39.6%
128	32.6%	30.7%	30.9%	29.1%	31.7%	36.3%	55.2%

4.3.3 Recognition Results

After phoneme sets normalization for MSA and ECA, it is possible to use MSA acoustic models instead of ECA acoustic models. The entire MSA corpus was used to train the acoustic model with a varying number of tied-states (from 62 to 4000) and Gaussian densities (from 1 to 128).

The speech recognition accuracy was evaluated based on batch decoding of the ECA testing set that consists of eight different speakers. A diagram that illustrates the usage of MSA in decoding ECA speech is shown in Fig. 4.4. The recognition accuracy in terms of WER is shown in Table 4.4. The optimized number of tied-states and the number of Gaussians per state are 1000 and 32 respectively. The absolute WER was found to be 24.5% with a +82.8% relative increase compared to the ECA phoneme-based baseline. The increase in WER was expected, as according to our assumption, the initial MSA acoustic model is considered as a dialect-independent model.

As we have seen, acoustic models trained with MSA speech data can be used to recognize dialectal Arabic speech. This approach may be very useful when dialectal speech resources are not available. In fact, this is the case for the majority of the Arabic dialects where transcribed speech resources are very limited.

4.4 Data Pooling Acoustic Modeling

In (Kirchhoff and Vergyri 2005), a cross-lingual approach was proposed that is based on data pooling. A data pool of MSA and ECA speech data was used to train the acoustic model. A digram that clarifies the concept is shown in Fig. 4.5. We have re-applied the same approach of data pooling on our speech data sets. We have created a speech data pool that consists of the post-processed MSA corpus (~33 hours) in conjunction with the ECA corpus (~0.5 hours). The data pool was

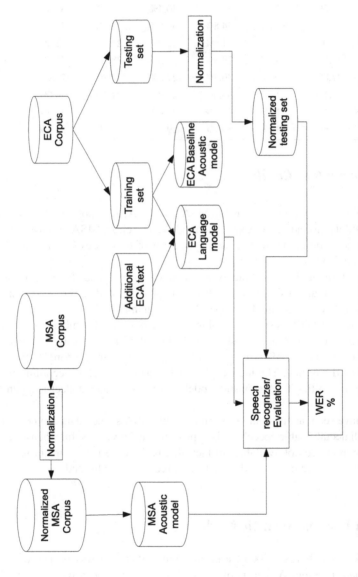

Fig. 4.4 System diagram for the proposed multilingual usage of phonemic MSA acoustic models for dialectal Arabic speech recognition

Fig. 4.5 Cross-lingual acoustic modeling using the data pooling approach to train an improved ECA acoustic model using ECA and MSA speech data

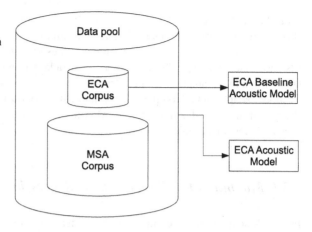

Table 4.5 Decoding results of the ECA test set using MSA and different phoneme-based acoustic model adaptation techniques

Acoustic model	WER	Relative
ECA	13.4%	baseline
MSA	24.5%	+82.8%
MSA/ECA Data Pool	21.4%	+59.7%
MSA/ECA MLLR	15.1%	+12.7%
MSA/ECA MAP	9.1%	−32.1%
MSA/ECA MLLR + MAP	7.8%	−41.8%

used to train a cross-lingual acoustic model. Finally, speech recognition results have indicated almost no reduction in WER compared to the ECA phonemic baseline. The WER was even increased as shown in Table 4.5 from 13.4% to 21.4% with an relative increase of 59.7%. In our case, the MSA data amount was almost 66 times the amount of dialectal data. Hence, the MSA data has completely masked all the acoustic features that exist in a relatively small amount of ECA speech data. Our interpretation is that the more the data of MSA, the less contribution of ECA. Statistically, a small ECA corpus will not be sufficient to change the acoustic model parameters (means, variances, mixture weights, and transition weights) in conjunction with a huge amount of MSA data which will have the dominant effect. Hence, the model will always be biased to MSA and adding more MSA data will decrease the effect of the ECA data.

4.5 Supervised Cross-Lingual Acoustic Modeling

We aim to benefit from a small amount of dialectal speech data that is phonetically transcribed and to benefit as well from existing MSA speech resources. Our approach described in this section is to first train an acoustic model using MSA data, then adapt this model with the samll amount of dialectal speech data using state-of-the-art acoustic model adaptation techniques.

In general, acoustic model adaptation assumes that an initial acoustic model exists and some adaptation data is available. The adaptation is called supervised adaptation when the adaptation data are transcribed. There exist two well known acoustic model adaptation techniques: Maximum Likelihood Linear Regression (MLLR) (Leggetter and Woodland 1995) and Maximum A-Posteriori (MAP) re-estimation (Lee and Gauvain 1993).

4.5.1 Maximum Likelihood Linear Regression (MLLR)

In MLLR adaptation, we compute a set of linear transformations for acoustic model parameters (mainly Gaussian means) of the Gaussian mixture model (GMM). The effect of these transformations is to shift the component means in the initial system so that each state in the HMM system is more likely to generate the adaptation data (Leggetter and Woodland 1995). In MLLR, all the adaptation data are used to estimate the transformation matrix for the acoustic model parameters. That is why MLLR is very suitable for unsupervised adaptation where the transcription can be noisy due to speech recognition errors. The parameters are transformed as:

$$\Phi_{MLLR} = A\Phi + b \tag{4.2}$$

Where Φ is a mean vector in the model and Φ_{MLLR} is the adapted mean using MLLR. (A, b) are the transformation parameters.

4.5.2 Maximum A-Posteriori (MAP)

Acoustic model adaptation may be performed using the Maximum A-Posteriori (MAP) re-estimation approach. This adaptation technique is also known as Bayesian adaptation. MAP benefits from the prior knowledge about the initial model parameters distribution (Lee and Gauvain 1993). Limited adaptation data would modify the initial model parameters with the guidance of prior knowledge. This is done to prevent large modifications of the initial parameters unless a sufficient amount of adaptation data is available.

$$\Phi_{MAP} = argmax_{\Phi} P(O|\Phi)P(\Phi) \tag{4.3}$$

Where Φ is the initial model parameter and Φ_{MAP} is the adapted parameter using MAP.

Usually MAP outperforms MLLR for supervised adaptation when sufficient adaptation data is available. In MAP, the adaptation data should be accurately transcribed. That is why MAP is not effective in unsupervised adaptation. One obvious drawback of the MAP adaptation is that it requires more adaptation data to be effective when compared to MLLR, because MAP adaptation is specifically defined at

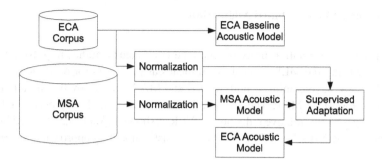

Fig. 4.6 Block diagram for phonemic cross-lingual acoustic modeling to build an improved ECA acoustic model

the component level. When larger amounts of adaptation training data become available, MAP begins to perform better than MLLR, due to this detailed update of each component (rather than the pooled Gaussian transformation approach of MLLR).

4.5.3 Adaptation Results

Basically, we are trying to adapt the MSA acoustic model, in order to make it dialect-dependent and hence improve recognition rate. The MSA acoustic model was adapted using the ECA training set along with the normalized transcriptions as shown in Fig. 4.6.

Three acoustic modeling adaptation techniques were evaluated:

MLLR Adaptation

Two iterations of MLLR were applied on the Gaussian means. In each iteration, the Gaussian means were transformed offline using the MLLR matrix. The adapted model resulted in 15.1% absolute WER with −38.4% absolute reduction compared to MSA alone, and we were able to get closer to the baseline accuracy by +12.7% relative increase in WER as shown in Table 4.5.

MAP Adaptation

MAP adaptation was found to give best results when adapting all acoustic model parameters: Gaussian means, variances, mixture weights, and transition weights. The absolute WER was 9.1% and it outperformed the baseline by −32.1% relative reduction in WER as shown in Table 4.5.

MLLR and MAP Combined Adaptation

In fact the two adaptation processes may be combined to futher improve performance, by using the MLLR transformed means as the priors for MAP adaptation. Therefore, we have combined both MLLR and MAP, starting by two iterations of MLLR followed by MAP. The absolute WER was 7.8% and it outperformed the baseline by −41.8% relative reduction in WER as shown in Table 4.5. The full block diagram for the cross-lingual MLLR/MAP acoustic modeling approach is shown in Fig. 4.7.

4.6 Effect of MSA Speech Amount on Supervised Adaptation

According to our assumption, MSA is always a second language for all Arabic speakers. That is why with more MSA speech data, we can train better dialectal Arabic acoustic models because of the broader coverage of the acoustic features for the different dialects. In this section, we are studying the effect of the amount of MSA speech data on the proposed phonemic MSA-based cross-lingual acoustic modeling approaches for dialectal Arabic.

We have started by taking only 0.5 hours of MSA speech data and we have repeated all the batch tests as shown earlier in Sections 4.3 and 4.5. Afterwards, all tests have been repeated using a doubled amount of MSA speech data. This process has been iterated until the entire MSA corpus has been used. It becomes obvious from Fig. 4.8 that while using MSA acoustic model as a multilingual model, the trend shows a consistent decrease in WER by using more MSA speech data. However, a breakeven with the ECA baseline did not occur. From this trend, we expect a breakeven if much more MSA data is added.

The same consistent decrease in WER has been also observed in MLLR adaptation. However, still a breakeven with the ECA baseline did not occur. For MAP adaptation, a breakeven was observed when an amount of only two hours of MSA data was used. For the combination of MLLR and MAP, a breakeven was observed very quickly when an amount of only one hour of MSA data was used.

4.7 Unsupervised Adaptation

Unsupervised adaptation is used when the adaptation speech data is not transcribed. It attempts first to guess the transcription of the adaptation data by using a speech recognizer. Then the estimated transcriptions along with adaptation data are used in acoustic model adaptation.

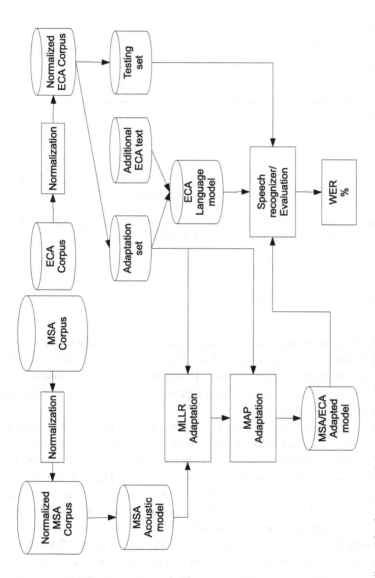

Fig. 4.7 Block diagram for the proposed phonemic cross-lingual MLLR/MAP acoustic modeling approach to build an improved ECA acoustic model

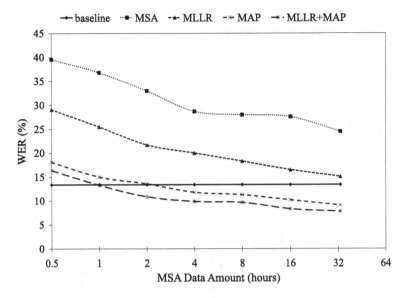

Fig. 4.8 Word Error Rate (WER) (%) for the different phonemic acoustic model supervised adaptation techniques while varying the amount of MSA speech data

4.7.1 Adaptation Approach

The proposed unsupervised adaptation approach assumes that only dialectal speech data is available without any transcriptions. We also assume that we do not have any language model for dialectal Arabic. That is why we initially planned to use the MSA corpus transcriptions to build an n-gram language model to automatically transcribe the ECA adaptation set.

Different n-gram language models were built (tri-gram, bi-gram, and uni-gram) using words as the lexical units. We also propose using phonemes as the lexical units to build the MSA language model. Different n-gram orders were used to build the phoneme-based language model (uni-gram, bi-gram, and tri-gram). A third type of language modeling is suggested which is the *flat uni-gram*. In the flat uni-gram language model, all the words (or phonemes) are assigned with an equal weight.

In order to estimate the transcriptions of the ECA adaptation set, the MSA acoustic model with the MSA language model were used to recognize the ECA adaptation set. The estimated transcriptions with the ECA adaptation set were finally used to adapt the initial MSA acoustic model using MLLR. MLLR adaptation was applied on Gaussians means of the model. Gaussian means were offline transformed using the MLLR matrix. Finally, to evaluate the proposed technique, the adapted acoustic model with the ECA language model was used to decode the ECA testing set and the WER was calculated.

4.7.2 Baseline and System Description

The block diagram for the proposed unsupervised adaptation approach is shown in Fig. 4.9. After the phoneme sets normalization, the entire MSA corpus was used to train the MSA acoustic model with a typical number of tied-states and Gaussians per state of 1000 and 32 respectively. The MSA acoustic model is the baseline in this case. The transcriptions of the ECA adaptation set were used to build a statistical bi-gram language model. The MSA acoustic model with the ECA language model was used to decode the ECA testing set. The baseline system resulted in an absolute WER of 24.5% as shown in Table 4.8.

4.7.3 Adaptation Results

Results show that we have a varying relative reduction in WER from −22.4% to −34.3% depending on the language modeling technique that was used in transcribing the adaptation set. Results are shown in Tables 4.6 and 4.7. It was noticed that the average WER of word-based language models is 18.2% with an average relative reduction in WER of −25.7%. Phoneme-based language models resulted in slightly better WER with an average of 17.8% and an average relative reduction in WER of −27.3%. Best results were found when applying the phoneme-based bi-gram model. This resulted in WER of 16.1% with a relative WER reduction of −34.3%. Flat uni-grams resulted in a better WER of 18.1% with a relative reduction of −26.1%. We have also tried using MAP adaptation, but the results were disappointing. MAP adaptation even resulted in an average of +15.0% relative increase in WER. The interpretation is that MAP is not suitable in unsupervised adaptation when the transcriptions of the adaptation set are fairly noisy.

In terms of adaptation speed, the Real Time Factor (RTF) was calculated while performing the automatic transcription of the adaptation set.

$$RTF = \frac{t_P}{t_I} \qquad (4.4)$$

Where t_P is the processing time (recognition time) and t_I is the total audio time.

The average RTF for word-based language models was found to be 2.0. For the phoneme-based language models, the RTF was found to be 0.9 with a relative reduction of −55.0% compared to the word-based approach.

4.8 Effect of MSA Speech Data Size on Unsupervised Adaptation

In this section, we study the effect of the MSA speech data size on unsupervised adaptation for dialectal Arabic. As we saw earlier, unsupervised adaptation can be very helpful for the Arabic dialects if phonetically transcribed speech corpora—required for accurate acoustic modeling—are not available. In order to study the

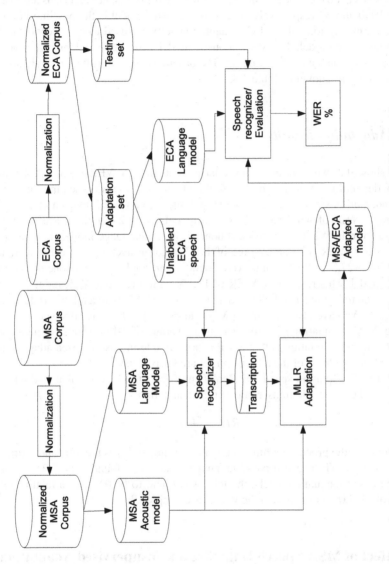

Fig. 4.9 System diagram for the proposed unsupervised phonemic cross-lingual adaptation system for dialectal Arabic

Table 4.6 Unsupervised acoustic model adaptation results in terms of Word Error Rate (WER) for different n-gram language models using words as the lexical units

Approach	WER	Relative
MSA (no adapt.)	24.5%	Baseline
Words flat uni-gram	18.1%	−26.1%
Words uni-gram	18.5%	−24.5%
Words bi-gram	18.3%	−25.3%
Words tri-gram	18.0%	−26.5%
Average	18.2%	−25.7%

Table 4.7 Unsupervised acoustic model adaptation results in terms of Word Error Rate (WER) for different n-gram language models using phonemes as the lexical units

Approach	WER	Relative
MSA (no adapt.)	24.5%	Baseline
Phonemes flat uni-gram	18.1%	−26.1%
Phonemes uni-gram	18.1%	−26.1%
Phonemes bi-gram	16.1%	−34.3%
Phonemes tri-gram	19.0%	−22.4%
Average	17.8%	−27.3%

Table 4.8 Adaptation data recognition speed in terms of real time factor (RTF) using different language modeling approaches

N-gram order	Lexical unit	
	Word	Phonemes
Flat uni-gram	3.2	1.0
Uni-gram	1.2	0.9
Bi-gram	1.8	0.9
Tri-gram	2.0	1.0
Average	2.0	0.9

effect of the MSA speech data size, we have repeated the experiment as in Section 4.6. The results are shown in Fig. 4.10 where they are compared against the ECA baseline. From Fig. 4.10, we may conclude that there is a consistent decrease in WER when more MSA speech data is added.

4.9 Summary

In order to improve the speech recognition rate for dialectal Arabic, it was shown in this chapter that we can benefit from existing MSA speech resources using state-of-the-art acoustic model adaptation techniques. A cross-lingual phonemic acoustic

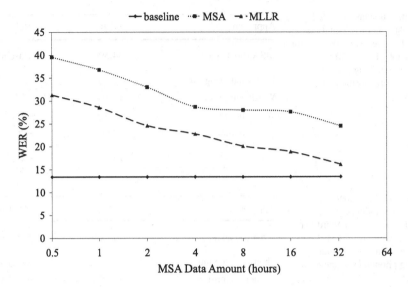

Fig. 4.10 Word Error Rate (WER) (%) for the phonemic acoustic model unsupervised adaptation using MLLR while varying the amount of MSA speech data

modeling approach has been proposed. It is based on supervised and unsupervised phonemic acoustic model adaptation. Egyptian Colloquial Arabic (ECA) has been chosen as a typical dialectal form. We have collected our own ECA speech corpus. It was divided into adaptation and testing sets. Several MSA acoustic models were trained using news broadcast speech. MSA acoustic models were adapted using MLLR and MAP with a little amount of ECA speech data. To make phoneme-based adaptation feasible, we have normalized the phoneme sets of MSA and ECA. Evaluation results show that the adapted MSA acoustic models outperform those trained with only ECA data. The proposed cross-lingual acoustic modeling approach showed best results when combining MLLR and MAP adaptations. In the case of supervised phoneme-based acoustic modeling, the adapted MSA model outperforms the ECA baseline by −41.8% relative reduction in WER.

Since for some Arabic dialect, phonetically transcribed speech resources may be not available, unsupervised adaption was also studied. The ECA adaptation set was automatically transcribed using a MSA acoustic model. The MSA acoustic model was then adapted using MLLR with the ECA adaptation set along with the recognized transcriptions. Recognition results show that we can achieve −22.4% to −34.3% relative reduction in WER compared to MSA as a single method.

The effect of the amount of MSA speech data has been studied. A consistent decrease in WER was observed while adding more MSA data. The consistent decrease in WER was observed with all proposed techniques including supervised and unsupervised adaptation.

In this chapter, we have seen how it is possible to improve a phonemic acoustic model for dialectal Arabic using existing MSA speech data sets. The proposed approach requires a relatively small amount of dialectal speech data. This data has to be phonetically transcribed in order to apply the proposed approach. Unfortunately, for the majority of the Arabic dialects, it is not always possible to phonetically transcribe dialectal data. In this case, the proposed approach of phonemic cross-lingual acoustic modeling cannot be applied. In the next chapter, we will propose a different approach where we can create acoustic models for the Arabic dialects for which speech data exist but without phonetic transcriptions.

Chapter 5
Graphemic Acoustic Modeling

5.1 Introduction

For many Arabic dialects, it is not always possible to estimate phonetic transcriptions to train a phonemic model. For these dialects, it is not possible to apply the acoustic modeling approach proposed in the previous chapter. In this chapter, we tackle the problem of having only graphemic transcriptions instead of phonemic ones. The problem is tackled by adopting grapheme-based acoustic modeling.

Our grapheme-based acoustic modeling (also known as graphemic modeling) for Arabic is an acoustic modeling approach where the phonetic transcription is approximated to be the word letters instead of the exact phonemes sequence. Short vowels and gemination can only be estimated from a fully diacritized Arabic script.

In Arabic phoneme-based acoustic modeling, we cannot just use a simple lookup table for phonetic transcription because of the morphological complexity nature of Arabic and the high homograph rate. We found that there is an average of 1.6 pronunciation variants for each word in MSA. The Arabic script does not usually include diacritic marks. That is why, in graphemic modeling, we rely on the acoustic model to implicitly model missing diacritics and all pronunciation variants. The main advantage of graphemic modeling is the rapid transcription development. Moreover, there is no need for manual or automatic diacritization, as in (Vergyri and Kirchhoff 2004; Sarikaya et al. 2006; Gal 2002; Habash and Rambow 2007; Nelken and Shieber 2005).

In Graphemic acoustic modeling, each grapheme (all Arabic letters except diacritics) are mapped to one single phoneme. Graphemic acoustic modeling is applied when the Arabic transcriptions do not contain diacritic marks. Phonemes that are estimated from diacritics are ignored as they can be implicitly modeled in the acoustic model as shown in Fig. 5.1.

A grapheme-based approach was introduced in (Billa et al. 2002) using MSA speech data since the transcriptions of the used corpus were mainly non-diacritized. It was noticed that the accuracy was acceptable. However, the results were not compared versus the phoneme-based approach. In (Billa et al. 2002), results with MSA show that the phoneme-based approach performs significantly better by a 14% absolute increase in accuracy compared to the grapheme-based approach. In order to

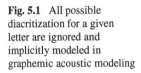

Fig. 5.1 All possible diacritization for a given letter are ignored and implicitly modeled in graphemic acoustic modeling

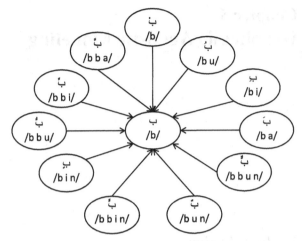

improve the performance of the grapheme-based approach, Lamel et al. (2007) proposed an explicit modeling for diacritic marks by using a generic vowel to replace all short vowels and creating all possible word pronunciation variants for any given non-diacritized word.

In this chapter, graphemic acoustic modeling is applied on dialectal Arabic speech data. We have started by experimenting the effect of different parameters on speech recognition accuracy in MSA. Afterwards, cross-lingual graphemic modeling was applied for dialectal Arabic where we can benefit from both existing large MSA data and some little dialectal Arabic data in a similar way to the phonemic acoustic modeling in Chapter 4.

5.2 Graphemic Modeling for MSA

We have started by calculating the frequency of the different diacritic marks in the Nemlar MSA speech corpus. There are nine types of diacritic marks in Arabic. Every written letter can be followed by one or two of them. We have calculated the frequency of each diacritic in the corpus and have found out that 44.9% of the corpus are diacritic marks as shown in Table 5.1. Consequently, in the grapheme-based (non-diacritized) transcription, we are missing around 44.9% of the information about the exact phonetic transcription. For example, the word /ta?'allama/ (he learned) contains five phonemes estimated from diacritics (four Fatha /a/ and one Shadda: doubling the consonant /l/), after removing the diacritics, the word will become /t?'lm/ (grapheme-based transcription). In this specific example, we are missing 55% of the information about the original phonetic transcription.

The frequencies of the different diacritic marks are shown in Table 5.2 where it was noticed that Fatha is the most frequently used diacritic with a frequency of 17.88%.

Table 5.1 Frequency of graphemes and diacritics in the diacritized transcriptions of a 33 hours MSA speech data (Nemlar news broadcast corpus)

Type	Frequency
Graphemes	1,179,623 (55.1%)
Diacritics	963,008 (44.9%)
Total	2,142,631 (100%)

Table 5.2 Arabic diacritics and their frequency of occurrence. The Tanween (Nunation) may only appear on the last letter of a word

Diacritic	Frequency
Fatha /a/	17.88%
Kasra /i/	9.98%
Damma /u/	3.53%
Shadda (consonant doubling)	3.40%
Sukun (no vowel)	9.49%
Tanween Fatha /an/	0.29%
Tanween Kasra /in/	0.33%
Tanween Damma /un/	0.07%

5.2.1 Graphemic Lexicon

The graphemic convention in our work is assigning one distinct phoneme for each letter except all forms of Alef and Hamza, as they were all assigned the same phoneme. Overall we have used 30 phonemes in our grapheme-based transcription convention.

The lexicon size in the graphemic system has been decreased by 37%. This is due to the absence of diacritic marks. Thus, all pronunciation variants for a given word (different diacritic combinations in the phoneme-based system) have been reduced to only one single pronunciation.

5.2.2 Assumptions and Procedure

In order to improve speech recognition accuracy by adopting graphemic acoustic modeling, our assumption is that an increased numbers of Gaussians per HMM states can implicitly model all diacritization possibilities for the same grapheme without the need of the full phonetic transcription. Furthermore, the grapheme-based acoustic modeling approach is much improved by increasing the amount of training data that are required to train the increased number of acoustic model parameters.

We assume also that by increasing the number of Gaussians or the amount of training data, the rate of improvement in the grapheme-based approach should be much higher than the rate of improvement in the phoneme-based approach. In other

Table 5.3 Lexicon size and average variants per word in the grapheme-based and the phoneme-based transcription systems (Nemlar news broadcast corpus)

	Lexicon size	Variants per word
Grapheme-based	39.2 K	1.0
Phoneme-based	62.7 K	1.6

words, the gap in the accuracy between the two approaches can be decreased by increasing any of the two parameters: the number of Gaussian densities or the amount of training data.

In order to prove our assumption, 28 hours ($\sim$85%) of the post-processed MSA corpus was taken as the training set and four hours ($\sim$15%) was taken as the testing set. Two transcription systems have been prepared phoneme-based transcription and grapheme-based transcription.

Phoneme-Based Transcription System

This system is the original phonetic transcription of the MSA corpus. The transcription contains all diacritic marks and hence the exact phonetic transcription is available. This system will be used in building the phoneme-based acoustic models. There are 34 phonemes in total (28 consonants, three long vowels, and three short vowels). Foreign and rare phonemes were ignored (/p/, /v/, /g/, and /l'/) and we mapped them to the closest standard Arabic phonemes. The lexicon size in this system is 62K words and in average 1.6 variants per word as shown in Table 5.3.

Grapheme-Based Transcription System

This is a grapheme only transcription system (letters without diacritics). All diacritic marks have been removed from the original transcription. Hence, this system contains only letters in the common Arabic writing system. Every letter is mapped to one phoneme and the system will be usedto build the grapheme-based acoustic models. In the grapheme-based transcription system the total number of unique phonemes that can be estimated from the text is 31 phonemes, the short vowels /a/, /i/, and /u/ are not included since they can be only estimated from the diacritics.

5.2.3 System Description

Training and decoding are based on CMU Sphinx engine. The number of states per HMM is three without skip state topology. We found that increasing the number of states per HMM or using skip state topology did not improve the accuracy in our experiment. All acoustic models are context-dependent tri-phone models with a

Table 5.4 Effect of Gaussian densities on PER in the grapheme-based approach (GBA) and the phoneme-based approach (PBA) using 28 hours of training data

Gaussian per state	PER (GBA)	PER (PBA)	Delta
1	49.76%	36.33%	13.43%
2	43.58%	32.37%	11.21%
4	37.37%	29.85%	7.52%
8	33.83%	27.97%	5.86%
16	30.63%	27.31%	3.32%
32	28.41%	26.33%	2.08%

total number of 2000 tied-states and 13 MFCC coefficients with 40 Mel frequency bands. The sampling rate is 16 kHz as in the original data.

One of the common problems in Arabic ASR is the high out-of-vocabulary (OOV) rate. For example, using a typical 65K lexicon in the domain of news broadcast, the OOV rate is 4% while in English it is less than 1% (Billa et al. 2002). In order to avoid this high OOV rate in Arabic, we decided to choosed phonemes as the lexical units. We have built a closed vocabulary 7-gram statistical language model using the CMU SLM toolkit (CMU 2010b) on the phoneme level by considering every phoneme as a word.

5.2.4 Effect of Gaussians per State

The entire 28 hours of the MSA training set has been used to train several acoustic models. All training and decoding parameters were fixed except the number of Gaussian densities. We started by using one Gaussian till reaching 32 Gaussians. The training was performed for all Gaussians using the grapheme-based transcription. Then it was repeated using the phoneme-based transcription. Overall six acoustic models were trained using the grapheme-based transcription and six models on the phoneme-based transcription. The entire four hours of the MSA testing set has been used in decoding and testing. The test was repeated for each acoustic model and the phoneme error rate (PER) was computed each time. Decoding results are shown in Table 5.4 and Fig. 5.2.

Results show that the accuracy can be improved by increasing the number of Gaussian densities in both the grapheme-based and phoneme-based approaches. However, the improvement rate is not the same. The rate of improvement in the grapheme-based approach was found to be higher than in the phoneme-based approach. In the case of one Gaussian, the absolute difference (Delta) in the accuracy between the two approaches was found to be 13.43%. The delta was found to decrease by increasing the number of Gaussians till reaching a delta of ~2% in the case of 32 Gaussians (see Fig. 5.3). In our experiment we found that by doubling the number of Gaussians, the delta is reduced in average by 30% relative reduction (using 28 hours of training data).

Fig. 5.2 Phoneme Error Rate
(PER) (%) versus the number
of Gaussian densities in the
grapheme-based approach
(GBA) and the
phoneme-based approach
(PBA) using 28 hours of
training data

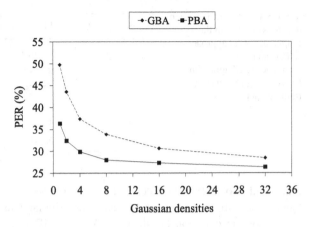

Fig. 5.3 Delta Phoneme
Error Rate PER(GBA-PBA)
(%) versus the number of
Gaussian densities using 28
hours of training data.

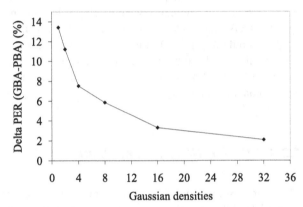

5.2.5 Effect of Training Data Amount

In order to study the effect of training data amount, the number of Gaussians per
state was fixed to 32. All training and decoding parameters kept unchanged except
the amount of training data. Using the grapheme-based transcription system, four
acoustic models have been created with different training data sizes (7, 14, 21, and
28 hours). Then, the training was repeated using the phoneme-based transcription
system. Overall four acoustic models were created using the grapheme-based transcription system and four acoustic models using the phoneme-based transcription
system. The entire four hours of the MSA testing set was used for the decoding. The
tests have been repeated for each acoustic model and the PER has been calculated
each time. Decoding results are shown in Table 5.5 and Fig. 5.4.

The results show that the accuracy could be improved by increasing the amount
of training data in both approaches. However, the rate of improvement is not
the same. It was found to be higher in the grapheme-based approach than in the
phoneme-based approach. By using seven hours of training data, the delta between
the accuracy in the two approaches was 9.25%. The delta in accuracy was found to

Table 5.5 Effect of the training data amount on PER in the phoneme-based approach (PBA) and the grapheme-based approach (GBA) using 32 Gaussians

Data amount (hrs)	PER (GPA)	PER (PBA)	Delta
7	49.45%	40.20%	9.25%
14	35.96%	28.82%	7.14%
21	32.18%	28.53%	3.65%
28	28.41%	26.33%	2.08%

Fig. 5.4 Phoneme Error Rate (PER) (%) versus the amount of training data in the grapheme-based approach (GBA) and the phoneme-based approach (PBA) using 32 Gaussians

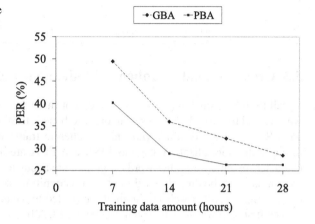

Fig. 5.5 Delta Phoneme Error Rate PER(GBA-PBA) (%) versus the amount of training data using 32 Gaussians

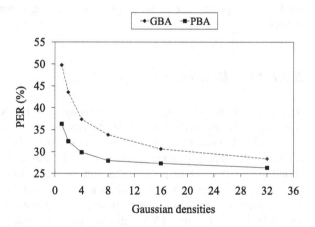

decrease by increasing the amount of training data till reaching a delta of 2.08% by using the whole 28 hours of the training set (see Fig. 5.5). In our experiment we found that the delta is reduced in average by 53% by doubling the amount of training data (using 32 Gaussians). This rate was not observed using eight Gaussians and below. The interpretation is that sufficient number of Gaussians should exist in order to notice a convergence between the accuracy of two approaches.

Table 5.6 Word Error Rate (WER) % while optimizing the total number of tied-states (TS) and the number of Gaussian densities for the grapheme-based ECA baseline acoustic models

Gaussians	Tied-states (TS)				
	62	125	250	500	1000
1	40.9%	33.3%	27.6%	25.9%	35.4%
2	32.2%	26.1%	21.5%	24.4%	34.3%
4	26.5%	20.6%	19.2%	23.8%	36.5%
8	23.2%	**18.7%**	20.0%	30.8%	49.1%
16	22.2%	18.8%	26.6%	49.6%	67.8%
32	24.6%	26.4%	41.7%	76.1%	90.0%

5.3 Cross-Lingual Graphemic Modeling for ECA

Recall from Section 4.3, MSA and ECA do not share the same phoneme sets, that is why we had to normalize both sets in order to benefit from the existing MSA speech data. Since we are dealing with only graphemic transcriptions, there is no need to apply the normalization stage as MSA and ECA are already sharing the same character inventory. In other words, a graphemic acoustic model trained with MSA speech data can be directly used in speech recognition for ECA. And according to our assumption, all vowels are implicitly modeled in the acoustic model.

The first step was to establish a graphemic ECA baseline system and evaluating speech recognition accuracy. Afterwards, the same proposed approaches in Chapter 4 are re-applied but with the graphemic modeling concept rather than phonemic modeling.

5.3.1 ECA Baseline

The grapheme-based baseline was built using only the ECA training set in a similar way to the phoneme-based baseline described in Section 4.2.2. However the lexicon is only graphemic in this case. It is prepared the same way as described in Section 5.2.1. The optimized numbers of tied-states and Gaussians per state were found to be 125 and 8 respectively. Search was done in the range from 62 to 1000 tied-states and from 1 to 32 Gaussians as shown in Table 5.6. Decoding the ECA test set using the baseline acoustic model resulted in an absolute WER of 18.7% as shown in Table 5.6 with a +5.3% absolute increase compared to the phoneme-based baseline (cf. Section 4.2.2).

5.3.2 MSA as a Multilingual Acoustic Model

The graphemic MSA acoustic model was trained in an analogous way to the phoneme-based experiment (cf. Section 4.3) except that the graphemic transcrip-

Table 5.7 Word Error Rate (WER) % while optimizing the total number of tied-states (TS) and the number of Gaussian densities using grapheme-based MSA acoustic models

Gaussians	Tied-states (TS)						
	62	125	250	500	1000	2000	4000
1	58.0%	54.3%	55.2%	56.7%	55.2%	59.0%	59.1%
2	52.4%	54.6%	47.8%	52.8%	53.7%	57.0%	57.8%
4	47.6%	47.6%	45.2%	43.7%	47.8%	53.3%	51.1%
8	50.0%	43.0%	44.1%	42.2%	43.3%	50.0%	49.8%
16	42.8%	40.0%	39.6%	38.7%	40.7%	47.2%	47.0%
32	42.0%	37.2%	36.3%	39.1%	40.4%	45.4%	50.9%
64	38.9%	**34.1%**	35.7%	39.6%	40.0%	44.8%	55.4%
128	39.6%	34.6%	38.5%	48.5%	40.7%	48.7%	67.2%

tion convention was applied. The entire MSA corpus was used to train the MSA acoustic model with a varying number of tied-states from 62 to 4000 and a varying number of Gaussians per state from 1 to 128.

The speech recognition accuracy was evaluated based on batch decoding of the ECA testing set that consists of eight different speakers. Fig. 5.6 shows a diagram that illustrates the usage of MSA to decode ECA (note that the normalization stage is eliminated in comparison to Fig. 4.4). The recognition accuracy in terms of WER is shown in Table 5.7. The optimized number of tied-states and the number of Gaussians per state were found to be 125 and 64 respectively.

The absolute WER was found to be 34.1% with a +82.4% relative increase compared to the baseline. The high WER increase is interpreted mainly due to the fact that ECA transcriptions do not follow the correct spelling as in MSA.

5.3.3 Data Pooling Acoustic Modeling

The data pooling approach that was proposed in (Kirchhoff and Vergyri 2005) was re-applied to build a graphemic acoustic model for ECA. Data pooling was adopted as illustrated earlier in Section 4.4 except that the transcription is only graphemic this time. Speech recognition results have indicated almost no reduction in WER compared to the ECA graphemic baseline. The WER increased as shown in Table 5.8 from 18.7% to 30.6% with a relative value of 63.6%.

5.3.4 Supervised Cross-Lingual Acoustic Modeling

The same adaptation techniques as described in Section 4.5 were re-applied here and the results are shown in Table 5.8. The block diagram for the cross-lingual

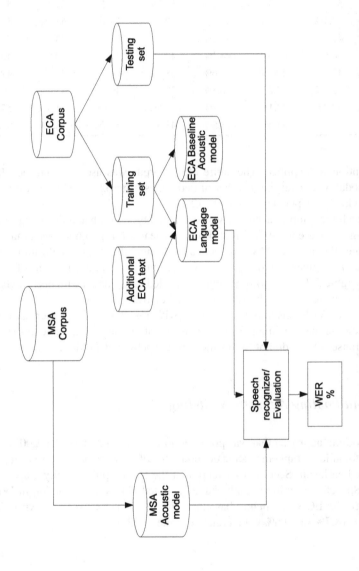

Fig. 5.6 System diagram for the proposed multilingual usage of graphemic MSA acoustic models for dialectal Arabic speech recognition

Table 5.8 Decoding results of the ECA test set using grapheme-based MSA acoustic model and the different adaptation techniques

Acoustic model	WER	Relative
ECA	18.7%	baseline
MSA	34.1%	+82.4%
MSA/ECA Data Pool	30.6%	+63.6%
MSA/ECA MLLR (unsupervised)	24.3%	+29.9%
MSA/ECA MLLR	23.4%	+25.1%
MSA/ECA MAP	19.0%	+1.6%
MSA/ECA MLLR + MAP	15.6%	−16.6%

graphemic adaptation system is shown in Fig. 5.7 where the normalization stage is removed. It was found that with MLLR adaptation, the absolute WER was 23.4% which is lower by −10.7% absolute compared to MSA alone but it is still higher than the baseline by +25.1% relative. With MAP adaptation, the absolute WER was 19.0% and it performed almost the same as the baseline with a negligible −1.2% relative reduction. When combining MLLR and MAP, the absolute WER was 15.6% and it outperformed the baseline by −16.6% relative reduction.

5.3.5 Unsupervised Graphemic Acoustic Modeling

In the case of unsupervised graphemic adaptation, we assume that dialectal speech data are available but without transcriptions. In an analogous way to what was described in Section 4.7, a bi-gram language model was trained using the MSA corpus transcriptions. Lexical units in this case are letters rather than phonemes. The MSA acoustic model and the letter-based bi-gram language model were used to automatically transcribe the ECA training set. The estimated ECA transcriptions were used to adapt the MSA acoustic model using MLLR adaptation as shown in Fig. 5.8. The approach resulted in 24.3% absolute WER with −28.7% relative decrease compared to a model trained with only MSA speech data.

5.3.6 Spelling Variants

Dialectal Arabic is mainly spoken and not formally written. That is why there is no single standard graphemic form for the same word. In many cases, we have different orthographic forms for a given word. Furthermore, written dialectal Arabic is highly affected by MSA, and writers tend to use the MSA graphemic forms instead of the form that matches the pronunciation. On the other hand, in MSA, transcriptions and pronunciations always match.

In other words, the transcription convention for the MSA corpus is certainly not the same as for dialectal Arabic although both corpora share the same character set.

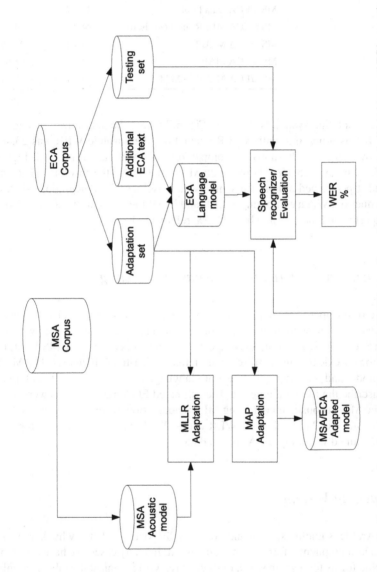

Fig. 5.7 Block diagram for the proposed graphemic cross-lingual MLLR/MAP acoustic modeling approach to build an improved ECA acoustic model

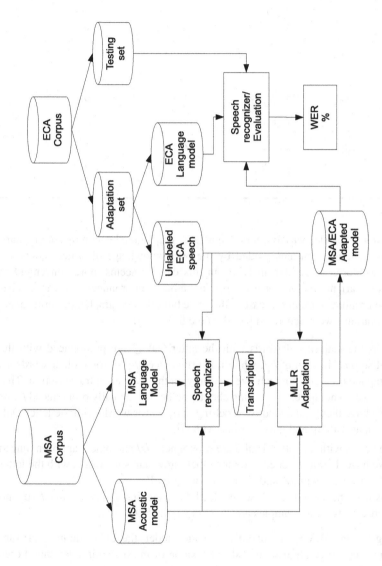

Fig. 5.8 System diagram for the proposed unsupervised graphemic cross-lingual adaptation system for dialectal Arabic

Table 5.9 Samples from the ECA graphemic lexicon after adding spelling variants, showing which variant is commonly written and which one matches pronunciation

Arabic	English	Spelling variants	Common written	Match pronunciation
ذرة	corn	ذ ر ة	✓	
		ز ر ة		
		د ر ة		✓
قهوة	coffee	ق ه و ة	✓	
		أ ه و ة		✓
قرية	village	ق ر ي ة	✓	✓
		أ ر ي ة		
ثوم	garlic	ث و م	✓	
		ت و م		✓
		س و م		

Therefore, our idea was to improve our ECA graphemic lexicon by adding automatically generated variants guided by the prior knowledge of which letters are usually not correctly written in ECA. In this case, it seems more convenient to name these variants *spelling* or *orthographic* rather than pronunciation variants because the pronunciation in this case is the same but different graphemic forms exist. Spelling variants were generated based on the following:

- In ECA the majority of words with the letter Qaf ق are pronounced with the glottal stop /ʔ/ instead of /q/, while it is pronounced /q/ in some other words. In the grapheme-based approach, we rely only on the graphemic transcription. That is why, we do not have information about whether it is actually pronounced /q/ or /ʔ/. We have therefore added two spelling variants for words with the letter Qaf: one variant with /q/ and the other one with /ʔ/.
- Some words with the letter Thal ذ are pronounced /z/ and others are pronounced /d/. We have therefore added two more spelling variants for words with the letter Thal: one variant with /z/ and the other one with /d/.
- Words with the letter Theh ث were added three spelling variants with /s/, /t/, and /θ/ respectively. Some samples are shown in Table 5.9.

Using the initial MSA graphemic acoustic model, the ECA training set transcriptions were force-aligned against our lexicon of multi-spelling variants. Force alignment should reduce spelling errors within the ECA graphemic transcriptions. More precisely, ECA transcriptions should now better represent the actual pronunciation compared to the initial ECA transcriptions. The force-aligned transcriptions along with the ECA training set were used to train new cross-lingual acoustic models based on all approaches previously described in Sections 5.3.3, 5.3.4, and 5.3.5.

Table 5.10 Effect of adding spelling variants on decoding results of the ECA test set using MSA and the different grapheme-based acoustic model adaptation techniques

Acoustic model	WER	Relative
ECA	18.7%	baseline
MSA	33.7%	+80.2%
MSA/ECA Data Pool	30.2%	+61.5%
MSA/ECA MLLR (unsupervised)	24.0%	+28.3%
MSA/ECA MLLR	23.9%	+27.8%
MSA/ECA MAP	17.3%	−7.5%
MSA/ECA MLLR + MAP	14.5%	−22.5%

The speech recognition results show that there is a significant reduction in WER compared to the case of not using spelling variants as shown in Table 5.10.

The initial MSA with the multi-spelling lexicon resulted in 33.7% absolute WER with −1.2% absolute reduction compared to the case of not using spelling variants. MLLR adaptation resulted in 23.9% absolute WER. MAP and MLLR + MAP outperformed the baseline by −7.5% and −22.5% relative reduction in WER respectively, and by −8.9% and −7.6% relative reduction compared to the same settings but without using spelling variants. It was noticed that when combining MLLR and MAP in the graphemic adaptation, the 14.5% WER is very close to the ECA phoneme-based baseline.

5.4   Cross-Lingual Graphemic Modeling for LCA

In order to prove that there is a consistent improvement in speech recognition accuracy across different Arabic dialects, we have re-applied the same proposed cross-lingual adaptation approach on Levantine Colloquial Arabic (LCA). Since the LCA corpus (Section 3.4) is only transcribed in graphemic form, only graphemic acoustic modeling is possible in this case.

The LCA speech corpus has been divided into training and testing data. The training set consists of ∼65% of the corpus and the testing set consists of the remaining ∼35%. All the previously mentioned steps in this chapter have been re-applied on LCA rather than ECA. The language model was trained as a bi-gram model using the LCA training set. Since the LCA corpus is larger than the ECA corpus, two configuration of the training set have been prepared as follows:

- *lca-set-13*: Consists of 13 different speakers.
- *lca-set-106*: Consists of 106 different speakers (the whole training set).

The training sets were used to create the LCA baseline and to adapt existing MSA acoustic models to build the cross-lingual acoustic models. Speech recognition results of using the lca-set-13 set are shown in Table 5.11. Results indicate that we are still able to observe significant reduction in WER compared to the LCA baseline. The LCA baseline resulted in 45.4% WER and the combination of MLLR and MAP adaptations resulted in 38.2% WER with −15.9% relative decrease.

Table 5.11 Decoding results of the LCA testing set and the lca-set-13 training/adaptation set using grapheme-based MSA acoustic model and the different adaptation techniques

Acoustic model	WER	Relative
LCA	45.4%	baseline
MSA	64.6%	+41.6%
MSA/ECA Data Pool	60.5%	+33.3%
MSA/ECA MLLR (unsupervised)	57.4%	+26.4%
MSA/LCA MLLR	56.7%	+24.9%
MSA/LCA MAP	41.8%	−7.9%
MSA/LCA MLLR + MAP	38.2%	−15.9%

Table 5.12 Decoding results of the LCA testing set and the lca-set-106 training/adaptation set using grapheme-based MSA acoustic model and the different adaptation techniques

Acoustic model	WER	Relative
LCA	24.6%	baseline
MSA	64.6%	+162.6%
MSA/ECA Data Pool	32.1%	+30.5%
MSA/ECA MLLR (unsupervised)	57.2%	+132.5%
MSA/LCA MLLR	56.4%	+129.3%
MSA/LCA MAP	24.4%	−0.8%
MSA/LCA MLLR + MAP	24.1%	−2.0%

Speech recognition results for the lca-set-106 set are shown in Table 5.12. They indicate that we are still able to observe reduction in WER compared to the LCA baseline. However in this case the results are closer to the baseline. The LCA baseline resulted in 24.6% WER and the combination of MLLR and MAP adaptations resulted in 24.1% WER with −2% relative decrease. In the case of the lca-set-106 set, the LCA data required for training is already large enough to create an acoustic model. That is why the relative improvement in WER was small. Despite the large amount of the LCA data, we are still able to benefit from the existing MSA speech resources to further improve speech recognition accuracy.

5.5 Discussion

Grapheme-based acoustic modeling may be considered as a multi-accent approach for Arabic, because of its ability to implicitly model different pronunciations for the same grapheme or letter. In other words, different possible diacritics without the need to explicitly include different pronunciation variants for the same word in the lexicon. For instance, the word /jal?'ab/ (he plays) in MSA is transformed to the word /jil?'ab/ in Egyptian Arabic and the only difference between the two words is: the vowel /a/ is transformed to the vowel /i/. This transformation is found in all present tense verbs in Egyptian Arabic. In the case of the grapheme-based approach, we do not need to make any changes in order to deal with the same word in the Egyptian accent. On the other hand, in the phoneme-based approach we will have to modify our lexicon to add the new word /jil?'ab/ as it is considered as a different pronunciation variant from the existing one /jal?'ab/.

5.6 Summary

In this chapter, we have shown how graphemic acoustic models are trained for Arabic when non-diacritized transcriptions are available. Graphemic acoustic modeling is very helpful when it is hard or not possible at all to perform phonetic transcriptions. In this case, the transcription is approximated to be the graphemic letters rather than the actual pronunciation, relying on the assumption that all missing vowels are going to be implicitly modeled in the acoustic model using a larger number of Gaussians per state.

Since dialectal Arabic is mainly spoken and not formally written, graphemic transcriptions and the actual spelling usually do not match as in the case of MSA transcriptions. We showed that by adding automatically generated spelling variants to the graphemic lexicon, we can force align ECA transcriptions using an initial MSA model to choose the correct spelling that matches the actual pronunciation.

The proposed cross-lingual acoustic modeling approach was tested on two Arabic dialects: Egyptian Colloquial Arabic (ECA) and Levantine Colloquial Arabic (LCA). Best results were observed when combining MLLR and MAP adaptations. For ECA, we were able to outperform the ECA baseline by -22.5% relative reduction in WER. For LCA, we were able to outperform the baseline by -2.0% to -15.9% relative reduction in WER depending on the amount of LCA training/adaptation data.

In the proposed approach, orthographic transcriptions were written in traditional Arabic orthography. Diacritic marks are always missing in traditional Arabic orthography. That is why short vowels were ignored and assumed to be implicitly modeled in the acoustic model. In the next chapter, we are considering another orthographic transcription method that is based on the *Arabic Chat Alphabet* (ACA) rather than the traditional Arabic alphabet. The ACA is written using Latin letter instead of Arabic ones. We have noticed that ACA is usually written using short vowels that are missing in traditional Arabic orthography. Hence, acoustic model trained using ACA-based transcriptions can perform better than grapheme-based transcriptions.

Chapter 6
Phonetic Transcription Using the Arabic Chat Alphabet

6.1 Introduction

Basically, Arabic is a morphologically very rich language. That is why a simple lookup table for phonetic transcription—essential for acoustic modeling—is not appropriate because of the high out-of-vocabulary (OOV) rate. Furthermore, Arabic orthographic transcriptions are written without diacritics. Diacritics are essential to estimate short vowels, nunation, gemination, and silent letters. State of the art techniques for the MSA phonetic transcription usually require several phases. In the first phase, transcriptions are written without diacritics. Afterwards, automatic diacritization is performed to estimate missing diacritic marks (WER is 15%–25%) as in (Kirchhoff and Vergyri 2005) and (Sarikaya et al. 2006). Finally, the mapping from diacritized text to phonetic transcription is almost a one-to-one mapping.

Dialectal Arabic usually differs significantly from MSA to the extent that dialects are considered as totally different languages. That is why phonetic transcription techniques for MSA cannot be applied directly on dialectal Arabic. In order to avoid automatic or manual diacritization, graphemic acoustic modeling was proposed for MSA (Billa et al. 2002) where the phonetic transcription is approximated to be the sequence of word letters while ignoring short vowels. It could be noticed that these graphemic systems work with an acceptable recognition rate. However, the performance is still below the accuracy of phonemic models. Since MSA and the Arabic dialects share the same character inventory, the grapheme-based approach was also applicable for dialectal Arabic as shown in (Vergyri et al. 2005) and (Elmahdy et al. 2010).

In this chapter, we propose the *Arabic Chat Alphabet* (ACA) for dialectal Arabic speech transcription instead of traditional Arabic orthography for the purpose of acoustic modeling. Transcriptions are written directly in ACA exactly as they appear in everyday life without any special training for transcribers. In fact, we have noticed that ACA is usually written with the short vowels that are omitted in normal Arabic orthography. Furthermore, it is a natural language that is very well known among all Arabic computer users. It was also found that the majority of computer users types in ACA faster than traditional Arabic orthography. Therefore, our assumption is that

M. Elmahdy et al., *Novel Techniques for Dialectal Arabic Speech Recognition*,   71
DOI 10.1007/978-1-4614-1906-8_6, © Springer Science+Business Media New York 2012

ACA-based transcriptions are closer to the full phonetically transcribed text rather than graphemic transcriptions (Arabic letters). Moreover, the transcription time is significantly reduced. In previous work in (Canavan et al. 1997), a Romanization notation was proposed for the Arabic phonetic transcription. However the proposed technique needs specifically trained transcribers that seems to be too costly given a significant long development time. We would like to highlight that we are not proposing a notation as the Buckwalter transliteration (see Table A.1) (Buckwalter 2002a). This notation cannot be directly used by any transcriber. It is only used to transliterate Arabic graphemes and diacritics into standard ASCII characters that are fully reversible to the original Arabic letters.

Egyptian Colloquial Arabic (ECA) has been chosen in our work as a typical dialect. Unlike other Arabic dialects, as we mentioned before, ECA was mainly selected since there exist some pronunciation dictionaries or lexicons like (Kilany et al. 2002) and (Hinds and Badawi 2009). Thus, it is possible to build a phonemic baseline which can be used to evaluate the performance of any other proposed technique. The main phonetic characteristics of ECA when compared to MSA have been already summarized in Chapter 4.

6.2 The Arabic Chat Alphabet

The *Arabic Chat Alphabet (ACA)* (also known as: Arabizi, Arabish, Franco-Arab, or Franco) is a writing system for Arabic in which English letters are written instead of Arabic ones. Basically, it is an encoding system that represents every Arabic phoneme with the English letter that matches the same pronunciation (Yaghan 2008). Arabic phonemes that do not have an equivalent in English are replaced with numerals or some accent marks that are close in shape to the corresponding Arabic letter. ACA was originally introduced when the Arabic alphabet was not available on computer systems. ACA is widely used in chat rooms, SMS, social networks, and non-formal emails. Actually, the majority still prefers writing Arabic in ACA instead of the original Arabic alphabet even if the Arabic alphabet is supported since ACA is much faster for them.

By comparing ACA-based text to normal Arabic orthography, it could be noticed that short vowels are usually written in ACA. For example, the word لبن (milk) is only written in traditional Arabic with three letters (ل, ب, and ن) from which we can only estimate the three consonants /l/, /b/, and /n/ and the vowels in-between are missing. In ACA, on the other hand, the same word is commonly written *laban* where we have the same three consonants plus two short vowels of the type /a/. Thus, the ACA-based script provides us with more information about missing vowels compared to normal Arabic orthography.

In order to prove our first assumption that ACA-based transcription is easier and faster than Arabic orthography (non-diacritized script), we conducted a survey over 100 Arabic computer users, 86% of the users confirmed that they type faster using ACA, 9% do not feel a difference, and 5% type Arabic letters slightly faster than ACA. All users confirmed that it is almost impossible to type a correct fully diacritized Arabic text. In order to further prove our assumption, recently, some research has been done to help the vast majority of Arabic computer users who are not able to type directly in Arabic orthography as in (Google 2009) and (Microsoft 2009) where automatic ACA conversion back to Arabic letters was proposed.

6.3 Speech Corpus

The same ECA corpus as in Chapter 4 has been used. The initial ECA phoneme set consists of twenty nine consonants and twelve vowels. Tables 6.1 and 6.2 show ECA phonemes in IPA, SAMPA, ACA, and the corresponding Arabic letters. In ECA, some consonants may be represented by more than one unique Arabic letter like /t/ and /d/. Some Arabic letters are represented by more than one form in ACA notation, like the letters Khah خ and Ghain غ . Short vowels are represented in Arabic by diacritic marks which is very uncommon to write them as the reader infers them from the context.

6.3.1 ACA Transcription Layer

Another transcription layer in ACA was added to the corpus. Since ACA may be variable across users, one transcriber does not seem to be able to catch all possible variations. Five different volunteer transcribers were asked to transcribe utterances from the corpus in ACA exactly as in their everyday life without any specific phonetic guidelines. The only requirement was to transcribe numbers in normalized word form rather than in digits. Fig. 6.1 shows some sample utterances with corresponding IPA and ACA transcriptions.

6.4 System Description and Baselines

The work on the ACA-based phonetic transcriptions relies on the same system as in Chapter 4 which is a GMM-HMM architecture based on the CMU Sphinx engine. The acoustic models are all fully continuous density context-dependent tri-phones with three states per HMM. A bi-gram language model with Kneser-Ney smoothing (Ney et al. 1994) was trained using the transcriptions of the ECA training set plus additional 25000 more utterances from the same speech domains. All language modeling parameters were fixed, so that any change in recognition rate is mainly due to acoustic modeling.

We have built two baseline systems for the two well-known acoustic modeling techniques for Arabic phoneme-based and grapheme-based (cf. Chapter 4 and Chapter 5).

Table 6.1 ECA consonants in IPA, SAMPA, and ACA notations with corresponding Arabic letters

IPA	Description	Arabic	SAMPA	ACA
ʔ	Plosive, voiceless glottal	ء (ق)	?	2, '
b	Plosive, voiced bilabial	ب	b	b
p	Plosive, voiceless bilabial	ب	p	p
t	Plosive, voiceless dental plain	ت (ث)	t	t
g	Plosive, voiced velar	ج	g	g, j
ʒ	Fricative, voiced postalveolar	ج	Z	j
ħ	Fricative, voiceless pharyngeal	ح	X\	7
x	Fricative, voiceless velar	خ	x	kh, 5, 7'
d	Plosive, voiced dental plain	د (ذ)	d	d
r	Trill, alveolar	ر	r	r
z	Fricative, voiced alveolar plain	ز (ذ)	z	z
s	Fricative, voiceless alveolar plain	س (ث)	s	s
ʃ	Fricative, voiceless postalveolar	ش	S	sh
ş	Fricative, voiceless alveolar emphatic	ص	s'	s, 9
ḍ	Plosive, voiced dental emphatic	ض (ظ)	d'	d, 9'
ţ	Plosive, voiceless dental emphatic	ط	t'	t, 6
ⱬ	Fricative, voiced alveolar emphatic	ظ (ض)	D'	z, 6'
ʕ	Fricative, voiced pharyngeal	ع	?\	3
ɣ	Fricative, voiced velar	غ	G	gh, 3'
f	Fricative, voiceless labio-dental	ف	f	f
v	Fricative, voiced labio-dental	ف	v	v
q	Plosive, voiceless uvular	ق	q	q, 8, 9
k	Plosive, voiceless velar	ك	k	k
l	Liquid, dental plain	ل	l	l
m	Nasal, bilabial	م	m	m
n	Nasal, alveolar	ن	n	n
h	Fricative, voiceless glottal	ه	h	h
w	Approximant (semi vowel), bilabial	و	w	w
j	Approximant (semi vowel), palatal	ي	j	y

6.4.1 Phoneme-Based Baseline

An ECA phoneme-based acoustic model baseline was trained with the ECA training set. The optimized number of Gaussians and tied-states were found to be 250 and 4 respectively. No approximations were applied on the phoneme set that consists of 41 phonemes. Speech recognition results of decoding the ECA test set using the baseline acoustic model were an absolute WER of 13.4% as shown in Table 6.3.

Table 6.2 ECA vowels in IPA, SAMPA, and ACA notations with corresponding Arabic diacritic marks. For each vowel there exist two forms: short and long. Long vowels are almost double the duration of the short ones

IPA	Description	Arabic	SAMPA	ACA
a	Low/open and front	ﹷ	a	a
ɑ	Low/open and back	ﹷ	A	a
i	High/close and front	ﹻ	i	i, e
e	Middle/half close and front	ﹻ	e	i, e
u	High/close and back	ﹹ	u	u, o
o	Middle/half close and back	ﹹ	o	u, o

Arabic:	يوم الجمعة ٢ أكتوبر ٢٠١٠
English:	Friday 2nd of October 2010
IPA:	/yoːm iggumʕa ʔitneːn ʔuktoːbɑr ʔalfeːn wʕɑʃɑːrɑ/
ACA:	/yoom iggum3a 2itneen 2uktoobar 2alfeen w3ashaara/
ACAST:	/yoom iggum3a 2itneen 2uktooba_r 2alfeen w3a_shaa_ra_/

Arabic:	تذكرة درجة أولى
English:	First class ticket
IPA:	/tɑzkɑrɑ dɑrɑgɑ ʔuːla/
ACA:	/tazkara daraga 2uula/
ACAST:	/ta_zka_ra_ da_ra_ga_ 2uula/

Arabic:	بيض مسلوق
English:	Boiled eggs
IPA:	/beːḍ masluːʔ/
ACA:	/bee9' masluu2/
ACAST:	/beed_ masluu2/

Fig. 6.1 Samples from the ECA speech corpus with Arabic, IPA, ACA, and the proposed ACAST transcriptions. Note that the shown ACA is the most correct form as ACA varies across transcribers

6.4.2 Grapheme-Based Baseline

Grapheme-based acoustic modeling (also known as graphemic modeling), as shown in Chapter 5, is an acoustic modeling approach for Arabic where the phonetic transcription is approximated to be the Arabic word letters instead of the exact phonemes sequence. All possible vowels, geminations, or nunations are assumed to be implicitly modeled in the acoustic model. The graphemic convention in our work is assigning one distinct phoneme to each letter except all forms of Alef and Hamza, these were all assigned the same phoneme model. The grapheme-based baseline was built using only the ECA training set in a similar way to the phoneme-based baseline but the lexicon in this case is only graphemic. The optimized number of Gaussians and tied-states were found to be 125 and 8 respectively. The result of decoding the ECA test set using the grapheme-based acoustic model was an absolute WER of

Table 6.3 Speech recognition results using phoneme-based, grapheme-based, and ACA-based acoustic models

Acoustic model	WER (%)	Relative1 (%)	Relative2 (%)
Phoneme-based	13.4	Baseline1	−28.0
Grapheme-based	18.6	+38.8	Baseline2
ACA-based *initial*	16.9	+26.1	−9.1
ACA-based *final*	14.1	+5.2	−24.2

18.6% as shown in Table 6.3 with a +38.8% relative increase in WER compared to the phoneme-based baseline.

6.5 ACA-Based Acoustic Modeling

6.5.1 Initial ACA-Based Model

Since there are different ACA representations for the same phoneme, we have normalized all ACA transcriptions of the ECA corpus into one consistent notation. More specifically, we have normalized the different phoneme representations to be mapped to the same phoneme model, e.g. /kh/, /5/, and /7'/ were all normalized to the same phoneme. Since the vowel /ɑ/ (emphatic /a/) does not have equivalent in ACA as it is always transcribed as /a/, it was removed from our phoneme set (long and short forms). Now our phoneme set consists of 39 phonemes. The normalized ACA transcriptions were used to train the initial ACA-based acoustic model. The optimized number of Gaussians and tied-states were found to be 125 and 8 respectively. An ACA-based lexicon was generated from the two parallel transcription layers: ACA and traditional Arabic. This lexicon was only used for testing in order to allow the same language model as in the baselines. The speech recognition result was an absolute WER of 16.9% as shown in Table 6.3 with a +26.1% relative increase in WER compared to the phoneme-based baseline, while it outperformed the grapheme-based baseline by −9.1% relative decrease in WER.

6.5.2 Modeling ACA Common Transcription Errors

ACA is not a standard encoding and it varies among transcribers. That is why we have summarized all common transcription errors that occurred in the ACA transcription layer of the corpus. The common errors (observed at least four times) were as follows:

1. The confusion between long and short vowels. For instant, using /a/ instead of /aa/ or vice versa.

2. The confusion between /y/, /i/, and /e/.
3. The confusion between /u/ and /o/.
4. Writing /ea/ instead of /ee/, e.g. instead of writing /2eeh/ (what), it is sometimes written /2eah/.
5. Ignoring the glottal stop /2/ at word beginnings, e.g. writing /ana/ (me) instead of /2ana/.
6. Geminations or consonant doublings are sometimes not written, e.g., writing /s/ instead of /ss/.
7. Ignoring /h/ at word endings.
8. The incorrect representation of emphatic phonemes like: /9/, /9'/, /6/, /6'/, and /q/ and typing the corresponding non-emphatic forms instead: /s/, /d/, /t/, /z/, and /k/ respectively.
9. Rare and foreign phonemes are sometimes miss represented, e.g. typing /g/ instead of /j/.

In order to model all ACA common errors, one idea was to generate all possible variants for a given word. However, this idea was quickly rejected since we found that the number of variants may be huge (may exceed 10 variants per word in average). The high number of variants increases confusability as the difference in pronunciation between words becomes smaller. Furthermore, it leads to a very complex search space as the decoder has to consider all of them.

Our alternative solution was to apply several approximations and phoneme-mergings to cover all error types (uncommon errors that occurred less than four times were not considered). This relies on the fact that they will be implicitly modeled in the acoustic model. ACA transcriptions were further pre-processed as follows:

1. Merging short vowels and their corresponding long vowels to the same phoneme model. In other words, regardless the transcription is /aa/ or /a/, both of them are normalized to /a/.
2. Merging /y/, /i/, and /e/.
3. Merging /u/ and /o/.
4. Normalizing /ea/ to /ee/.
5. Automatically adding /2/ to the beginning of words starting with vowels (in Arabic, we have a constraint that words or utterances cannot start with vowels).
6. Approximating double consonants by single consonants.
7. Deleting /h/ if found at word endings.
8. Merging emphatic consonants and their corresponding non-emphatic forms.
9. Normalizing foreign phonemes to the nearest Arabic ones, so /v/, /p/, and /j/ approximating them to /f/, /b/, and /g/ respectively.

Finally, the total number of phonemes was reduced to 23. To clarify the concept of normalization at this stage by an example, the words: /2eah/, /2eeh/, /eah/, /eeh/, /eih/, /ea/, etc (what) are all mapped to the same model.

6.5.3 Modeling ACA Ambiguity

Some phoneme sequences in ACA are phonetically ambiguous, e.g. the consonant /3'/ that corresponds to the Arabic letter Ghain can be written /gh/ in many cases. The question arises whether the sequence /gh/ represent /3'/ or /g/ followed by /h/. Other ambiguous examples are: /kh/ and /sh/ that may correspond the letters Khah and Sheen respectively. Our solution was to initially train a preliminary context-independent acoustic model. All possible pronunciation variants have been automatically generated for ambiguous sequences as long as they satisfy two constraints. Firstly, no more than two consecutive consonants are allowed in Arabic. Secondly, the allowed syllables in Arabic are CV, CVC, and CVCC where C is a consonant and V is a vowel. It should be noted that the CVCC pattern can only be found at word endings. Afterwards, the context-independent acoustic model is used to force align ACA transcriptions to select the most probable pronunciation variants. Finally, the forced aligned transcriptions are used to train the final ACA-based acoustic model.

6.5.4 ACA-Based Final Results

After ACA common errors and ambiguity modeling, speech recognition results indicate an absolute WER of 14.1% with a small relative increase of +5.2% compared to the phoneme-based baseline (see Table 6.3). This implies that ACA-based models can perform almost as accurate as phoneme-based models. This is mainly due to the representation of short vowels. It can also be noticed that the ACA-based model has outperformed the grapheme-based baseline by −24.2% relative decrease in WER which is a major improvement over graphemic modeling for dialectal Arabic, since accurate phonetic transcription for the different Arabic dialects, in most cases, is too costly and may be not feasible at all for some dialects. Results also confirm that all the applied approximations are all acceptable and contribute to a better recognition rate.

In our proposed approach of using ACA orthographic transcriptions, the ACA-based acoustic model has outperformed the grapheme-based one—given that transcription time was also reduced—because of several facts:

1. In grapheme-based modeling, short vowels are not explicitly modeled. While in ACA, the majority of short vowels are explicitly modeled. We have found that in Arabic in general, ~40% of the speech consists of vowels.
2. Moreover, traditional graphemic transcriptions for dialectal Arabic are highly affected by MSA. For example, writing the letter Qaf /q/ even though it is realized as a glottal stop /2/ in ECA. In ACA, on the other hand, MSA influence is not observed and transcriptions always follow the correct realized pronunciation.
3. In traditional Arabic alphabet, the letter Teh Marboota and the letter Alef Maksura are ambiguous. The letter Teh Marboota is either pronounced /h/, /a/ or /t/. The letter Alef Maksura is sometimes written instead of the letter Yeh, and hence

Table 6.4 ECA phonemes in IPA, SAMPA, and the proposed ACAST notations. For each vowel there exist two forms: short and long. Long vowels are almost double the duration of the short ones

IPA	SAMPA	ACAST	IPA	SAMPA	ACAST
ʔ	?	2	ɣ	G	gh
b	b	b	f	f	f
p	p	p	v	v	v
t	t	t	q	q	q
g	g	g	k	k	k
ʒ	Z	j	l	l	l
ħ	X\	7	m	m	m
x	x	kh	n	n	n
d	d	d	h	h	h_
r	r	r	w	w	w
z	z	z	j	j	y
s	s	s	a	a	a
ʃ	S	sh	ɑ	A	a_
ṣ	s‘	s_	i	i	i
ḍ	d‘	d_	e	e	e
ṭ	t‘	t_	u	u	u
ẓ	D‘	z_	o	o	o
ʕ	?\	3			

it is either pronounced /y/ or /a/. Such ambiguity is already resolved in ACA transcriptions.

4. In traditional Arabic alphabet, nunation at word endings is not usually written. In ACA, nunation is always correctly represented.

6.6 ACAST: A New Phonetic Notation for Arabic

We have proposed a new phonetic transcription notation for dialectal Arabic called *Arabic Chat Alphabet for Speech Transcription* (*ACAST*). The notation was intended to Egyptian Colloquial Arabic. However, it can be extended to cover the more dialects as well. Our motivation was to develop a phonetic notation that is readable by humans as well as machines. The SAMPA notation, on the other hand, is only readable by computers and it is rather difficult for humans to read or to review transcriptions written in SAMPA. The ACAST notation was basically derived from the common ACA notation and it is shown in Table 6.4. Some samples are shown in Fig. 6.1. We have carefully chosen the ACAST notation to satisfy the following criteria:

- Match as much as possible the most frequently used ACA forms.

- Can be directly integrated with popular speech recognition engines like CMU Sphinx (CMU 2010a) and HTK (Cambridge 2010) without the need of special converters as in the case of SAMPA notation where it cannot be directly used.
- Should be uniquely parsable. In other words, no ambiguity is expected when the symbols are written without spaces in-between.
- Can be extended to other Arabic varieties (not only ECA).

In ACAST, the underscore "_" symbol is used to differentiate between some emphatic phonemes and the corresponding non-emphatic ones. Furthermore, to avoid any possible ambiguity that may arise when writing concatenated text as in the case of /kh/, /gh/, /sh/, and /h_/.

6.7 Summary

In this chapter, we have proposed the Arabic Chat Alphabet (ACA) for dialectal Arabic speech transcription. Our assumption is that ACA is a natural language that includes short vowels that are missing in traditional Arabic orthography. Furthermore, the ACA transcriptions can be rapidly prepared. Egyptian Colloquial Arabic was chosen as a typical dialect. Two speech recognition baselines were built: phonemic and graphemic. Original transcriptions were re-written in ACA by different transcribers. Ambiguous ACA sequences were handled by automatically generating all possible variants. ACA variations across transcribers were modeled by phoneme normalization and merging. Results show that the ACA-based approach outperforms the graphemic baseline while it performs as accurate as the phoneme-based baseline with a slight increase in WER.

The ACA-based approach is very promising for rapid phonetic transcription development for dialectal Arabic where the Arabic Chat Alphabet is used exactly as in everyday life for the purpose of acoustic modeling. The proposed approach represents an alternative way to conventional phonetic transcription methods. Results show that ACA-based acoustic models trained with ACA transcriptions perform almost as accurate as the phoneme-based models with a small relative increase in WER of +5.2%. Furthermore, the ACA-based acoustic models were found to outperform grapheme-based models by −24.2% relative decrease in WER. This is mainly due to the lack of explicit modeling of short vowels in traditional Arabic orthography. Finally, the performance of the ACA-based models was mainly interpreted because of the appearance of short vowels in ACA transcripts. The proposed approach may be extended not only to the different Arabic varieties but to other languages as well where the chat alphabet is widely used like: Persian, Urdu, Hindi, Bengali, Hebrew, Greek, Serbian, Chinese, Russian, etc.

Finally, we have proposed a phonetic notation for Egyptian Colloquial Arabic that is called ACAST that is close as possible to ACA. The ACAST notation is supposed to be readable by humans and machines as well which is not possible using common notations including SAMPA and IPA.

Chapter 7
Conclusions and Future Directions

7.1 Summary

In our work, we addressed the problem of how to improve ASR for dialectal Arabic. Dialectal Arabic is only spoken and not formally written. There is no common standard for dialectal Arabic orthography. Moreover, dialectal Arabic differs significantly from the standard Arabic form which is known as *Modern Standard Arabic* (MSA).

Basically, high quality and large speech corpora are required to train ASR systems. However, the existing dialectal Arabic speech corpora are very sparse and are of low quality. For some Arabic dialects, speech resources do not exist at all. That is why it is difficult to train a reliable ASR system for dialectal Arabic.

With our research we mainly pursued three goals. The first one is to collect a high quality dialectal speech corpus. The second goal is to find a way to benefit from existing large corpora of MSA. The third goal is to rapidly develop phonetic transcriptions for dialectal speech data.

We have started with an overview about Arabic language from a speech recognition point of view. We have seen that there exist a lot of Arabic varieties. These were classified into Standard and Dialectal. A comparison of some Arabic dialects was provided including Modern Standard Arabic (MSA), Classical Arabic, and Egyptian Colloquial Arabic (ECA) (Elmahdy et al. 2009a).

Since, we could not find an existing phonetically labeled high quality speech corpus for dialectal Arabic; we have decided to collect our own speech corpus for ECA. A high quality recording step was established and all recordings were phonetically transcribed and reviewed (Elmahdy et al. 2010, 2011). Another corpus for Levantine Colloquial Arabic (LDC) was chosen from the Linguistic Data Consortium (LDC). However the LCA corpus is not phonetically labeled.

In order to improve the recognition rate of dialectal Arabic, it was shown that we can benefit from existing MSA speech resources using acoustic model adaptation techniques (Elmahdy et al. 2010, 2009c). A cross-lingual acoustic modeling approach has been proposed. It is based on supervised and unsupervised phonemic acoustic model adaptation. Egyptian Colloquial Arabic (ECA) has been chosen

M. Elmahdy et al., *Novel Techniques for Dialectal Arabic Speech Recognition*, 81
DOI 10.1007/978-1-4614-1906-8_7, © Springer Science+Business Media New York 2012

as a typical dialectal form. The ECA corpus was divided into adaptation and testing sets. Several MSA acoustic models were trained using news broadcast speech. MSA acoustic models were adapted using MLLR and MAP with our in-house collected ECA corpus. To make phoneme-based adaptation feasible, we have normalized the phoneme sets of MSA and ECA. Results show that the adapted MSA acoustic models outperform acoustic models trained with only ECA data. The proposed cross-lingual acoustic modeling approach showed best results when combining MLLR and MAP adaptations. In the case of supervised phoneme-based acoustic modeling, the adapted MSA model outperforms the ECA baseline by -41.8% relative reduction in WER (Elmahdy et al. 2010).

Since for some Arabic dialect, phonetically transcribed speech resources may be not available, unsupervised adaption was also studied. The ECA adaptation set was automatically transcribed using MSA acoustic models. The MSA acoustic model was then adapted using MLLR with the ECA adaptation set along with the recognized transcriptions. Recognition results show that we can achieve -22.4% to -34.3% relative reduction in WER compared to MSA alone.

The effect of the amount of MSA speech data has been studied. A consistent decrease in WER was observed while adding more MSA data. This decrease was observed with all proposed techniques including supervised and unsupervised adaptation.

We showed how graphemic acoustic models can be trained for Arabic when non-diacritized transcriptions are available. Graphemic acoustic modeling seems to be very helpful when phonetic transcription is hard or not possible to perform. In this case, the phonetic transcription is approximated to be the graphemic letters rather than the actual pronunciation. This relies on the fact that all the missing vowels are going to be implicitly modeled in the acoustic model using a larger number of Gaussians per state (Elmahdy et al. 2009b).

Since dialectal Arabic is mainly spoken and not formally written, graphemic transcriptions usually do not match the actual spelling as in the case of MSA transcriptions. We showed that by adding automatically generated spelling variants to the graphemic lexicon, we can force align ECA transcriptions using an initial MSA model to choose the correct spelling that matches the actual pronunciation (Elmahdy et al. 2010).

The proposed cross-lingual acoustic modeling approach was tested on two Arabic dialects: Egyptian Colloquial Arabic (ECA) and Levantine Colloquial Arabic (LCA). Best results were observed when MLLR and MAP adaptations were combined. For ECA, we were able to outperform the ECA baseline by -22.5% relative reduction in WER. For LCA, we were able to outperform the baseline by -2.0% to -15.9% relative reduction in WER depending on the amount of LCA training/adaptation data.

Finally, we proposed the Arabic Chat Alphabet (ACA) as a naturally everyday life alphabet for dialectal Arabic speech transcription. Our assumption is that ACA is a natural language that includes short vowels that are missing in traditional Arabic orthography. Furthermore, ACA transcriptions can be rapidly prepared. Two speech recognition baselines were built: phonemic and graphemic. Original transcriptions were re-written in ACA by different transcribers. Ambiguous ACA sequences were handled by automatically generating all possible variants. ACA variations across transcribers were modeled by phonemes normalization and merging. Results show that the ACA-based approach outperforms the graphemic baseline while it performs as accurate as the phoneme-based baseline with a slight increase in WER.

ACA-based transcription results are very promising for rapid phonetic transcription development for dialectal Arabic where the ACA is used exactly as in everyday life for the purpose of acoustic modeling. The proposed approach represented an alternative way to conventional phonetic transcription methods. Results show that ACA-based acoustic models trained with ACA transcriptions perform almost as accurate as the phoneme-based models with a small relative increase in WER of +5.2%. Furthermore, the ACA-based acoustic models were found to outperform grapheme-based models by −24.2% relative decrease in WER. This is mainly due to the lack of explicit modeling of short vowels in traditional Arabic orthography (Elmahdy et al. 2011). Finally, the performance of the ACA-based models was mainly interpreted because of the appearance of short vowels in ACA transcripts.

7.2 Contributions

Our contributions can be classified into three main categories: theoretical, practical, and experimental.

7.2.1 Theorectical

1. Cross-lingual acoustic modeling for dialectal Arabic was investigated where a large MSA corpus is used to train an initial phoneme-based acoustic model. The MSA acoustic model was then adapted with a relatively small amount of dialectal speech data.
2. For the Arab dialects where it is not possible to estimate accurate phonetic transcriptions, we proposed a graphemic acoustic modeling where phonetic transcriptions are approximated to be the word letters. Cross-lingual acoustic modeling was applied across MSA and dialectal Arabic using the graphemic modeling approach.
3. We have developed a novel approach where we can benefit from the *Arabic Chat Alphabet* (ACA) to phonetically transcribe dialectal Arabic speech corpora. The ACA is a natural language where Latin letters are used instead of Arabic ones. ACA text includes short vowels that are usually missing in Arabic orthography.

7.2.2 Practical

1. In our work, we have collected and phonetically labeled a high quality Egyptian Colloquial Arabic (ECA) corpus. It has been used as a reference to evaluate

all proposed approaches in our work for improving ASR accuracy for dialectal Arabic.
2. We have implemented a framework for dialectal Arabic ASR. The framework is based on the CMU-Sphinx engine where it was configured to train Arabic acoustic models and language models. The architecture was further modified to decode and recognize Arabic speech.
3. Since MSA and dialectal Arabic do not share the same phoneme sets, it was initially impossible to adapt existing MSA acoustic model with an amount of dialectal Arabic speech. In order to benefit from existing MSA speech corpora in training acoustic models for dialectal Arabic, we propose a phoneme set normalization to normalize the phoneme sets of MSA and dialectal Arabic into one set. Thus, it is possible to cross-lingually adapt acoustic models across MSA and dialectal Arabic.

7.2.3 Experimental

1. While evaluating the speech recognition accuracy for cross-lingual acoustic models, a significant improvement was observed in both phonemic and graphemic modeling. Recognition accuracy outperformed the performance of acoustic models trained with only dialectal speech data.
2. There was a consistent reduction in WER by adding more MSA data in the training of cross-lingual models. In other words, ASR accuracy for dialectal Arabic can be improved by only adding MSA speech data which is easily available.
3. In order to prove that the proposed approach can be extended to other Arabic dialects, the approach was repeated with Levantine colloquial Arabic (LCA). Significant improvement in ASR accuracy was also observed while using cross-lingual acoustic models trained with MSA and then adapted with LCA speech data.

7.3 Future Work

We are proposing different future directions in the area of dialectal Arabic speech recognition:

- The proposed approach of cross-lingual acoustic modeling may be extended to other Arabic dialects to check the variation in speech recognition accuracy improvements among the different dialects.
- Text resources for dialectal Arabic are also very limited and lack standardization. That is why training statistical language models for dialectal Arabic remains a problem since it requires large text corpora. Our idea is that there still exist some similarities between dialectal Arabic and MSA on the lexical, syntactic, and morphological level. Therefore, a novel way should be proposed to benefit

from existing MSA text resources in order to improve dialectal Arabic language modeling.

- Since there exist a large variety of Arabic dialects, one important task is to identify the Arabic dialect from the speech signal. Classification can be performed using a statistical classifier. The classifier has to be trained using as many dialects as possible.
- The proposed approach of using the Arabic Chat Alphabet (ACA) may be extended not only to the different Arabic varieties but to other languages as well where the chat alphabet is widely used like: Persian, Urdu, Hindi, Bengali, Hebrew, Greek, Serbian, Chinese, Russian, etc.

Appendix A
Buckwalter Transliteration

Table A.1 Arabic Letters with windows 1256, ISO 8859-6, and Unicode character encoding and corresponding Buckwater transliteration

Letter	Description	Win. CP-1256	ISO 8859-6	Unicode	Buckwalter
ء	Letter Hamza	C1	C1	U+0621	'
آ	Letter Alef, Madda above	C2	C2	U+0622	|
أ	Letter Alef, Hamza above	C3	C3	U+0623	>
ؤ	Letter Waw, Hamza above	C4	C4	U+0624	&
إ	Letter Alef, Hamza Below	C5	C5	U+0625	<
ئ	Letter Yeh, Hamza above	C6	C6	U+0626	}
ا	Letter Alef	C7	C7	U+0627	A
ب	Letter Beh	C8	C8	U+0628	b
ة	Letter Teh Marbuta	C9	C9	U+0629	p
ت	Letter Teh	CA	CA	U+062A	t
ث	Letter Theh	CB	CB	U+062B	v
ج	Letter Jeem	CC	CC	U+062C	j
ح	Letter Hah	CD	CD	U+062D	H
خ	Letter Khah	CE	CE	U+062E	x
د	Letter Dal	CF	CF	U+062F	d
ذ	Letter Thal	D0	D0	U+0630	*
ر	Letter Reh	D1	D1	U+0631	r
ز	Letter Zain	D2	D2	U+0632	z
س	Letter Seen	D3	D3	U+0633	s
ش	Letter Sheen	D4	D4	U+0634	$
ص	Letter Sad	D5	D5	U+0635	S
ض	Letter Dad	D6	D6	U+0636	D
ط	Letter Tah	D8	D7	U+0637	T

M. Elmahdy et al., *Novel Techniques for Dialectal Arabic Speech Recognition*,
DOI 10.1007/978-1-4614-1906-8, © Springer Science+Business Media New York 2012

Table A.1   (continued)

Letter	Description	Win. CP-1256	ISO 8859-6	Unicode	Buckwalter
ظ	Letter Zah	D9	D8	U+0638	Z
ع	Letter Ain	DA	D9	U+0639	E
غ	Letter Ghain	DB	DA	U+063A	g
_	Tatweel	DC	E0	U+0640	_
ف	Letter Feh	DD	E1	U+0641	f
ق	Letter Qaf	DE	E2	U+0642	q
ك	Letter Kaf	DF	E3	U+0643	k
ل	Letter Lam	E1	E4	U+0644	l
م	Letter Meem	E3	E5	U+0645	m
ن	Letter Noon	E4	E6	U+0646	n
ه	Letter Heh	E5	E7	U+0647	h
و	Letter Waw	E6	E8	U+0648	w
ى	Letter Alef Maksura	EC	E9	U+0649	Y
ي	Letter Yeh	ED	EA	U+064A	y
ً	Fathatan	F0	EB	U+064B	F
ٌ	Dammatan	F1	EC	U+064C	N
ٍ	Kasratan	F2	ED	U+064D	K
َ	Fatha	F3	EE	U+064E	a
ُ	Damma	F5	EF	U+064F	u
ِ	Kasra	F6	F0	U+0650	i
ّ	Shadda	F8	F1	U+0651	~
ْ	Sukun	FA	F2	U+0652	o
ٰ	Letter Superscript Alef	–	–	U+0670	`
ٱ	Letter Alef Wasla	–	–	U+0671	{
پ	Letter Peh	81	–	U+067E	P
چ	Letter Tcheh	8D	–	U+0686	J
ڤ	Letter Veh	–	–	U+06A4	V
گ	Letter Gaf	90	–	U+06AF	G

Appendix B
Egyptian Colloquial Arabic Lexicon

Table B.1 ECA corpus lexicon with Arabic othography, Buckwalter transliteration, and SAMPA phonetic transcription

Arabic	Buckwalter	SAMPA	Arabic	Buckwalter	SAMPA
اه	lh	?A:	ابجدي	>bjdy	?abgadi
ابدا	>bdA	?abadan	ابريل	>bryl	?abri:l
ابو	>bw	?abu	ابيض	>byD	?AbyAd'
اتاخر	<t<xr	it?AxxAr	اتفاق	<tfAq	?ittifa:?
اتناثر	<tnA$r	?itnA:SAr	اتنين	<tnyn	?itne:n
اثار	>vAr	?AsA:r	اجازة	>jAzp	?aga:za
احمر	>Hmr	?AX\mAr	احنا	<HnA	?iX\na
اخبارك	>xbArk	?AxbA:rAk	اخضر	>xDr	?Axd'Ar
اد	>d	?add	ادارة	<dArp	?idA:ra
اذاعة	<*AEp	?iza:?\a	اذن	<*n	?izn
اذنك	<*nk	?iznak	اراضي	>rADy	?ArA:d'i
اربعة	>rbEp	?ArbA?\a	اربعتاشر	>rbEtA$r	?ArbA?\tA:SAr
اربعين	>rbEyn	?arbi?\i:n	اربع	>rbE	?ArbA?\
ارض	>rD	?Ard'	ارنب	>rnb	?arnab
ازاي	<zAy	izza:y	ازرق	>zrq	?azra?
ازيك	<zyk	?izzayyak	اسانسير	>sAnsyr	?AsAnse:r
اسد	>sd	?asad	اسكندرية	<skndryp	?iskindiriyya
اسوان	>swAn	?AswA:n	اسود	<swd	?iswid
اسيوط	>sywT	?asyu:t	اشكال	>$kAl	?aSka:l
اشكرك	>$krk	aSkurak	اشكرك(٢)	>$krk(2)	?ASkurAk
اصفر	>Sfr	?As'fAr	اصل	>Sl	?As'l
اعلان	<ElAn	?i?\la:n	اعلانات	<ElAnAt	?i?\lA:na:t
اعمل	>Eml	?a?\mil	اعياد	>EyAd	?a?\ya:d

M. Elmahdy et al., *Novel Techniques for Dialectal Arabic Speech Recognition*,
DOI 10.1007/978-1-4614-1906-8, © Springer Science+Business Media New York 2012

Table B.1 (continued)

Arabic	Buckwalter	SAMPA	Arabic	Buckwalter	SAMPA
اغسطس	>gsTs	?aGust'us	افريقيا	>fryqyA	?afriqya
افغانستان	>fgAnstAn	?afGanista:n	اكتوبر	>ktwbr	?ukto:bAr
الا	<lA	illa	الابجدية	Al<bjdyp	il?abgadiyya
الاحمر	Al<Hmr	il?AX\mAr	الاربع	Al<rbE	lArbA?\
الاردن	Al<rdn	?il?urdun	الارضي	Al<rDy	il?Ard'i
الانفاق	Al<nfAq	l?anfa:?	الاتنين	Al<tnyn	litne:n
الاسعاف	Al<sEAf	il?is?\a:f	البحر	AlbHr	?ilbAX\r
البلد	Albld	ilbalad	التلات	AltlAt	ittala:t
الجزاير	AljzAyr	?iggaza:yir	الجمعة	AljmEp	iggum?\a
الحد	AlHd	ilX\add	الحنفية	AlHnfyp	ilX\anafiyya
الخميس	Alxmys	ilxami:s	الخير	Alxyr	ilxe:r
الدائري	AldA}ry	idda:?iri	الدسم	Aldsm	iddasam
الدم	Aldm	iddamm	الدور	Aldwr	?iddo:r
الدين	Aldyn	iddi:n	الساعة	AlsAEp	?issa:?\a
السبت	Alsbt	issabt	السلامة	AlslAmp	ssala:ma
السلام	AlslAm	?issala:mu	الشمس	Al$ms	iSams
الشنطة	Al$nTp	iSSAnt'a	الصحة	AlSHp	is's'iX\X\A
الطريق	AlTryq	?it't'Ari:?	الظهر	AlZhr	id'd'uhr
العربية	AlErbyp	il?\ArAbiyya	العموم	AlEmwm	l?\umu:m
الغردقة	Algrdqp	?ilGarda?a	الف	>lf	?alf
الف(٢)	>lf(2)	?alif	الفسحة	AlfsHp	ilfusX\a
الفين	>lfyn	?alfe:n	القاهرة	AlqAhrp	?ilqA:hirA
القطر	AlqTr	?il?At'r	الله	Allh	?AllA:
الله(٢)	Allh(2)	?AllA:h	اللي	Ally	lli
اللي(٢)	Ally(2)	?illi	الماسورة	AlmAswrp	?ilmAsu:rA
المانيا	>lmAnyA	?almanya	المرة	Almrp	?ilmArrA
المشكلة	Alm$klp	ilmuSkila	المغرب	Almgrb	?ilmaGrib
الملوك	Almlwk	lmulu:k	المواعين	AlmwAEyn	?ilmawa?\i:n
الموضوع	AlmwDwE	ilmAwd'u:?\	النهار	AlnhAr	innAhA:r
النهارده	AlnhArdh	innAhArdA	النور	Alnwr	innu:r
الهرم	Alhrm	?ilhArAm	الهول	Alhwl	lho:l
الوقت	Alwqt	ilwa?t	الوقفة	Alwqfp	ilwa?fa
اماكن	>mAkn	?ama:kin	امان	>mAn	?ama:n
امبارح	<mbArH	imba:riX\	امبارح(٢)	<mbArH(2)	?imba:riX\
امريكا	>mrykA	?amri:ka	انا	>nA	?ana
انبوبة	>nbwbp	?anbu:bit	اهل	>hl	?ahl
اهلا	>hlA	?ahlan	اهله	>hlh	?ahlu

Table B.1 (continued)

Arabic	Buckwalter	SAMPA	Arabic	Buckwalter	SAMPA
اوتوبيس	>wtwbys	?utubi:s	اوضة	>wDp	?o:d'it
اول	>wl	?awwil	اولى	>wlY	?u:la
اي	>y	?ayy	ايطاليا	<yTAlyA	?it'Alya
ايه	<yh	?e:	ايه(٢)	<yh(2)	?e:h
ايوه	>ywh	?aywa	باب	bAb	ba:b
بالك	bAlk	ba:lak	باللبن	bAllbn	billaban
بترول	btrwl	bitro:l	بتنجان	btnjAn	bitinga:n
بتوجعني	btwjEny	btiwga?\ni	برتقان	brtqAn	burtu?a:n
برسيم	brsym	barsi:m	برقوق	brqwq	bar?u:?
بزر	bzr	bizr	بسلة	bslp	bisilla
بصل	bSl	bAs'Al	بطاطس	bTATs	bAt'A:t'is
بطاطين	bTATyn	bAt'At'i:n	بطانية	bTAnyp	bAt't'Aniyya
بطن	bTn	bAt'n	بطني	bTny	bAt'ni
بطيخ	bTyx	bAt't'i:x	بعد	bEd	ba?\d
بعدين	bEdyn	ba?\de:n	بعض	bED	bA?\d
بقالك	bqAlk	ba?a:lak	بقالي	bqAly	ba?a:li
بقدونس	bqdwns	ba?du:nis	بقرة	bqrp	ba?ArA
بكره	bkrh	bukrA	بكرة	bkrp	bukrA
بلح	blH	balaX\	بلدي	bldy	baladi
بنات	bnAt	bana:t	بنت	bnt	bint
بنزينة	bnzynp	banzi:na	بنزين	bnzyn	banzi:n
بنطلون	bnTlwn	bAnt'Alo:n	بنعناع	bnEnAE	bini?\na:?\
به	bh	bih	بهارات	bhArAt	buhArA:t
بوابة	bwAbp	bawwa:ba	بو تجاز	bwtjAz	butaga:z
بودرة	bwdrp	budrA	بوري	bwry	bu:ri
بوسطة	bwsTp	bust'a	بوكية	bwkyp	buke:h
بيبسي	bybsy	pepsi	بيت	byt	be:t
بيتزا	bytzA	pitza	بيضا	byDA	be:d'a
بيض	byD	be:d'	بيضة	byDp	be:d'A
بينا	bynA	bi:na	تاخد	tAxd	ta:xud
تالاف	tAllf	tala:f	تاني	tAny	ta:ni
تحت	tHt	taX\t	تحليل	tHlyl	taX\li:l
تذكرة	t*krp	tAzkArA	ترايزة	trAbyzp	t'ArAbe:zA
ترايزة(٢)	trAbyzp(2)	t'ArAbe:zit	ترتيب	trtyb	tarti:b
ترعة	trEp	tir?\a	تسع	tsE	tisa?\
تسعة	tsEp	tis?\a	تسعتاشر	tsEtA$r	tisa?\tA:SAr

Table B.1   (continued)

Arabic	Buckwalter	SAMPA	Arabic	Buckwalter	SAMPA
تسعمية	tsEmyp	tus?\umiyya	تسعميت	tsEmyt	tus?\umi:t
تسعين	tsEyn	tis?\i:n	تشكيلة	t$kylp	taSki:la
تصبح	tSbH	tis'bAX\	تعبك	tEbk	ta?\abak
تعيش	tEy$	ti?\i:S	تفاح	tfAH	tuffa:X\
تفاحة	tfAHp	tuffa:X\a	تقريبا	tqrybA	ta?ri:ban
تلات	tlAt	talat	تلاتاشر	tlAtA$r	tAlAttA:SAr
تلاتة	tlAtp	tala:ta	تلاتين	tlAtyn	talati:n
تلت	tlt	tilt	تلتمية	tltmyp	tultumiyya
تلتميت	tltmyt	tultumi:t	تمانية	tmAnyp	tamanya
تمانين	tmAnyn	tamani:n	تمثيل	tmvyl	tamsi:l
تمساح	tmsAH	timsa:X\	تمن	tmn	taman
تمنتاشر	tmntA$r	tAmAntA:SAr	تنمية	tmnmyp	tumnumiyya
تمنميت	tmnmyt	tumnumi:t	تنساش	tnsA$	tinsa:S
تنس	tns	tinis	توصلني	twSlny	tiwAs's'Alni
تيل	tyl	ti:l	تيه	tyh	tih
ثه	vh	sih	جاذبية	jA*byp	ga:zibiyya
جايين	jAyyn	gayyi:n	جبر	jbr	gAbr
جبنة	jbnp	gibna	جديد	jdyd	gidi:d
جديدة	jdydp	gidi:da	جراج	jrAj	gArA:Z
جرالك	jrAlk	gArA:lak	جرجير	jrjyr	gargi:r
جرس	jrs	gArAs	جرنال	jrnAl	gurnA:l
جزار	jzAr	gAzza:r	جزر	jzr	gAzAr
جزمة	jzmp	gazma	جزيرة	jzyrp	gizi:ra
جعان	jEAn	ga?\a:n	جلطة	jlTp	gAlt'A
جمبري	jmbry	gambari	جمل	jml	gamal
جمهور	jmhwr	gumhu:r	جمهورية	jmhwryp	gumhuriyyit
جنية	jnyp	gine:	جوافة	jwAfp	gawa:fa
جواهرجي	jwAhrjy	gawahirgi	جورج	jwrj	ZorZ
جيش	jy$	ge:S	جيم	jym	gi:m
جينس	jyns	Zins	جيوش	jyw$	guyu:S
حاجة	HAjp	X\a:ga	حال	HAl	X\a:l
حتة	Htp	X\itta	حجز	Hjz	X\agz
حد	Hd	X\add	حداشر	HdA$r	X\idA:SAr
حروف	Hrwf	X\uru:f	حزام	HzAm	X\iza:m
حساب	HsAb	X\isa:b	حساسية	HsAsyp	X\asasiyya
حصان	HSAn	X\us'A:n	حصل	HSl	X\As'al
حفلة	Hflp	X\aflit	حق	Hq	X\a??
حلاق	HlAq	X\alla:?	حمار	HmAr	X\umA:r

Table B.1 (continued)

Arabic	Buckwalter	SAMPA	Arabic	Buckwalter	SAMPA
حمام	HmAm	X\amma:m	حمدله	Hmdllh	X\amdilla
حه	Hh	hAh	حوالين	HwAlyn	X\awa:le:n
حوالي	HwAly	X\awa:li	حوض	HwD	X\o:d
حيوانات	HywAnAt	X\ayawa:na:t	خاتم	xAtm	xa:tim
خدني	xdny	xudni	خرطوم	xrTwm	xart'u:m
خشب	x$b	xaSab	خصوصية	xSwSyp	xus'us'iyya
خضرا	xDrA	xAd'rA	خلاط	xlAT	xAllA:t
خلي	xly	xalli	خمس	xms	xamas
خمسة	xmsp	xamsa	خمستاشر	xmstA$r	xAmAstA:SAr
خمسمية	xmsmyp	xumsumiyya	خمسميت	xmsmyt	xumsumi:t
خمسين	xmsyn	xamsi:n	خه	xh	xAh
خوخ	xwx	xo:x	خيار	xyAr	xiyA:r
خير	xyr	xe:r	دا	dA	da
دال	dAl	da:l	دايمة	dAymp	dayma
درجة	drjp	dArAgA	دروس	drws	duru:s
دعوة	dEwp	da?\wa	دقايق	dqAyq	da?a:yi?
دقيقة	dqyqp	di?i:?a	دكتور	dktwr	dukto:r
دلوقتي	dlwqty	dilwa?ti	ديسمبر	dysmbr	disimbir
ذال	*Al	za:l	ذرة	*rp	dura
ذي	*y	zayy	راجل	rAjl	ra:gil
ربعمية	rbEmyp	rub?\umiyya	ربعميت	rbEmyt	rub?\umi:t
ربع	rbE	rub?\	ربنا	rbnA	rAbbina
رحلة	rHlp	riX\la	رخصة	rxSp	ruxs'it
رز	rz	ruzz	رصاص	rSAS	rus'A:s'
رغيف	rgyf	riGi:f	رمادي	rmAdy	rumA:di
ره	rh	rih	رومي	rwmy	ru:mi
رياضة	ryADp	riyA:d'A	زبدة	zbdp	zibda
زراعة	zrAEp	zirA:?\A	زراعي	zrAEy	zirA:?\i
زعلان	zElAn	za?\la:n	زيادة	zyAdp	ziya:da
زيارة	zyArp	ziyA:rA	زيارتك	zyArtk	ziyArtak
زيت	zyt	ze:t	زيتون	zytwn	zatu:n
زيرو	zyrw	zi:ru	زين	zyn	ze:n
سادة	sAdp	sa:da	ساقعة	sAqEp	sa??\a
سباحة	sbAHp	siba:X\a	سبانخ	sbAnx	saba:nix
سبب	sbb	sabab	سبتمبر	sbtmbr	sibtimbir
سبع	sbE	saba?\	سبعة	sbEp	sab?\a

Table B.1 (continued)

Arabic	Buckwalter	SAMPA	Arabic	Buckwalter	SAMPA
سبعتاشر	sbEtA$r	sABA?\tA:SAr	سبعمية	sbEmyp	sub?\umiyya
سبعميت	sbEmyt	sub?\umi:t	سبعين	sbEyn	sab?\i:n
ستاشر	stA$r	sittA:SAr	ست	st	sit
ستة	stp	sitta	ستمية	stmyp	suttumiyya
ستميت	stmyt	suttumi:t	ستين	styn	sitti:n
سجن	sjn	sign	سخنة	sxnp	suxna
سرة	srp	surrA	سعر	sEr	si?\r
سعيدة	sEydp	sa?\i:da	سفرة	sfrp	sufra
سفير	sfyr	safi:r	سكر	skr	sukkar
سلامتك	slAmtk	salamtak	سمحت	smHt	samaX\t
سمسم	smsm	simsim	سمك	smk	samak
سمكة	smkp	samaka	سنين	snyn	sini:n
سواق	swAq	sawwa:?	سواقة	swAqp	siwa:?a
سوداني	swdAny	suda:ni	سوريا	swryA	surya
سين	syn	si:n	شارع	$ArE	Sa:ri?\
شامي	$Amy	Sa:mi	شاي	$Ay	Sa:y
شجرة	$jrp	SAgarA	شفته	$fth	Suftu
شقة	$qp	Sa??a	شكر	$kr	Sukr
شكرا	$krA	SukrAn	شكل	$kl	Sakl
شمس	$ms	Sams	شمسية	$msyp	Samsiyya
شمعة	$mEp	Sam?\a	شنطة	$nTp	SAnt'A
شهادة	$hAdp	Siha:dit	شواية	$wAyp	Sawwa:ya
شوربة	$wrbp	Surba	شوكة	$wkp	So:ka
شوية	$wyp	Swayya	شين	$yn	Si:n
صابون	SAbwn	s'Abu:n	صاد	SAd	s'A:d
صالون	SAlwn	s'alo:n	صباح	SbAH	s'AbA:X\
صفر	Sfr	s'ifr	صندوق	Sndwq	sandu:?
صوت	Swt	s'o:t	ضاد	DAd	d'A:d
ضغط	DgT	d'AGt'	طبق	Tbq	t'AbA?
طبي	Tby	t'ibbi	طعمية	TEmyp	t'A?\miyya
طفاية	TfAyp	t'AffA:ya	طماطم	TmATm	t'AmA:t'im
طموح	TmwH	t'umu:X\	طه	Th	t'Ah
طوبة	Twbp	t'u:bA	طول	Twl	t'u:l
طيارة	TyArp	t'AyyA:rA	طيران	TyrAn	t'AyArA:n
ظه	Zh	D'Ah	عارف	EArf	?\a:rif
عاريحة	EAlryHp	?\arri:X\a	عالسلامة	EAlslAmp	?\assala:ma
عجيب	Ejyb	?\agi:b	عدس	Eds	?\ads

Table B.1   (continued)

Arabic	Buckwalter	SAMPA	Arabic	Buckwalter	SAMPA
عربي	Erby	?\ArAbi	عربيات	ErbyAt	?\ArAbiyya:t
عربية	Erbyp	?\ArAbiyya	(٢)عربية	Erbyp(2)	?\ArAbiyyit
عسل	Esl	?\asal	عشر	E$r	?\ASAr
عشرة	E$rp	?\ASA:rA	عشرين	E$ryn	?\iSri:n
عصافير	ESAfyr	?\As'Afi:r	عصفورة	ESfwrp	?\As'fu:rA
عصير	ESyr	?\As'i:r	عظام	EZAm	?\iD'A:m
عقلي	Eqly	?\a?li	علبة	Elbp	?\ilbit
على	ElY	?\ala	عليكو	Elykw	?\ale:ku
عمري	Emry	?\umri	عن	En	?\an
عنب	Enb	?\inab	عندك	Endk	?\andak
عنكبوت	Enkbwt	?\ankabu:t	عنوان	EnwAn	?\inwA:n
عيد	Eyd	?\i:d	عيش	Ey$	?\e:S
عين	Eyn	?\e:n	غروب	grwb	Guru:b
غير	gyr	Ge:r	غيرها	gyrhA	Gerha
غين	gyn	Ge:n	فاتت	fAtt	fa:tit
فاصوليا	fASwlyA	fAs'ulya	فاضل	fADl	fA:d'il
فاضية	fADyp	fAd'ya	فالقلب	fAlqlb	fil?alb
فبراير	fbrAyr	fibrA:yir	فراخ	frAx	fira:x
فرامل	frAml	fArA:mil	فرخة	frxp	farxa
فرصة	frSp	furs'A	فرنسا	frnsA	fArAnsA
فضة	fDp	fAd'd'A	فضلك	fDlk	fAd'lAk
فظيع	fZyE	faD'i:?\	فلافل	flAfl	fala:fil
فلفل	flfl	filfil	فلوس	flws	filu:s
فه	fh	fih	فودافون	fwdAfwn	vodafon
فوق	fwq	fo:?	فول	fwl	fu:l
في	fy	fi	فيزا	fyzA	vi:za
فيلا	fylA	villa	فيلم	fylm	film
فينو	fynw	fi:nu	قاف	qAf	qA:f
قانون	qAnwn	qAnu:n	قبل	qbl	?abl
قديم	qdym	?adi:m	قرار	qrAr	qArA:r
قرد	qrd	?ird	قرية	qryp	qAryA
قريش	qry$	?ari:S	قطة	qTp	?ut't'A
قطن	qTn	?ut'n	قطنة	qTnp	?ut'nA
قلب	qlb	?alb	قلم	qlm	?alam
قمح	qmH	?amX\	قمر	qmr	?AmAr
قهوة	qhwp	?ahwa	قيمة	qymp	?i:ma
كاف	kAf	ka:f	كام	kAm	ka:m

Table B.1 (continued)

Arabic	Buckwalter	SAMPA	Arabic	Buckwalter	SAMPA
كامل	kAml	ka:mil	كان	kAn	ka:n
كبدة	kbdp	kibda	كبريت	kbryt	kabri:t
كتاب	ktAb	kita:b	كتكوت	ktkwt	katku:t
كداب	kdAb	kadda:b	كده	kdh	kida
كدة	kdp	kida	كمان	kmAn	kama:n
كنبة	knbp	kanaba	كندا	kndA	kanada
كندوز	kndwz	kandu:z	كهرباء	khrbA'	kAhrAbA
كوبري	kwbry	kubri	كورة	kwrp	ko:rA
كوسة	kwsp	ko:sa	كيس	kys	ki:s
كيفك	kyfk	ke:fak	لا	l<	la?
الا	l<A	la?a	لازم	lAzm	la:zim
لام	lAm	la:m	لبن	lbn	laban
لحمة	lHmp	laX\ma	لقطة	lqTp	lu?t'a
لبة	lmbp	lAmbA	لو	lw	law
لوبيا	lwbyA	lubya	لوحة	lwHp	lo:X\a
ليا	lyA	liyya	ليلة	lylp	le:la
ليمون	lymwn	lamu:n	ما	mA	ma
مارس	mArs	ma:ris	مالك	mAlk	ma:lak
مالوش	mAlw$	malu:S	ماليش	mAly$	mali:S
مانجة	mAnjp	manga	مايو	mAyw	ma:yu
مترو	mtrw	mitru	متوسط	mtwsT	mutAwAssit
مجنون	mjnwn	magnu:n	مجهول	mjhwl	maghu:l
محاسب	mHAsb	muX\a:sib	محامي	mHAmy	muX\a:mi
محدش	mHd$	maX\addS	محطة	mHTp	mAX\At't'it
محمرة	mHmrp	miX\AmmAra	مخدرات	mxdrAt	muxAddArA:t
مدمس	mdms	midammis	مراية	mrAyp	mira:ya
مربوطة	mrbwTp	mArbu:t'A	مرة	mrp	mArrA
مرتب	mrtb	murattab	مساء	msA'	masa:?
مساعدة	msAEdp	musa?\da	مستشفى	mst$fY	mustaSfa
مستغرب	mstgrb	mistAGrAb	مسدودة	msdwdp	masdu:da
مسلوق	mslwq	maslu:?	مسموح	msmwH	masmu:X\
مش	m$	miS	مشمش	mm	miSmiS
مشوي	m$wy	maSwi	مشوية	m$wyp	maSwiyya
مصر	mSr	mAs'r	مضرب	mDrb	mAd'rAb
مطار	mTAr	mAt'A:r	مطبخ	mTbx	mAt'bAx
مطعم	mTEm	mAt'?\Am	مع	mE	ma?\a
معاك	mEAk	ma?\a:k	معايا	mEAyA	ma?\a:ya
معدنية	mEdnyp	ma?\daniyya	معقول	mEqwl	ma?\?u:l

Table B.1 (continued)

Arabic	Buckwalter	SAMPA	Arabic	Buckwalter	SAMPA
معلقة	mElqp	ma?\la?a	معلقة(٢)	mElqp(2)	ma?\la?it
مغامرات	mgAmrAt	muGAmrA:t	مفتاح	mftAH	mufta:X\
مفلفل	mflfl	mefalfel	مفيش	mfy$	mafi:S
مقلي	mqly	ma?li	مكان	mkAn	maka:n
مكرونة	mkrwnp	mAkAro:na	مكسوف	mkswf	maksu:f
مكوة	mkwp	makwa	مكوجي	mkwjy	makwagi
ملح	mlH	malX\	ممثلة	mmvlp	mumassila
ممكن	mmkn	mumkin	من	mn	min
منديل	mndyl	mandi:l	منطقة	mnTqp	mAnt'i?a
منها	mnhA	minha	مهندس	mhnds	muhandis
مواعيد	mwAEyd	mawa?\i:d	موجود	mwjwd	mawgu:d
موز	mwz	mo:z	موسيقى	mwsyqY	musi:qA
مية	myp	miyya	مية(٢)	myp(2)	mAyyA
ميت	myt	mi:t	ميتين	mytyn	mite:n
ميلاد	mylAd	mila:d	ميم	mym	mi:m
ناخد	nAxd	na:xud	نادي	nAdy	na:di
ناس	nAs	na:s	نايم	nAym	na:yim
نحل	nHl	naX\l	نشاط	n$AT	nASA:t
نصيحة	nSyHp	nAs'i:X\a	نظارة	nZArp	nAd'd'A:rA
نعم	nEm	na?\am	نعناع	nEnAE	ni?\na:?
نفس	nfs	nafs	نقطة	nqTp	nu?t'it
نمل	nml	naml	نملة	nmlp	namla
نوفمبر	nwfmbr	nuvimbir	نوم	nwm	no:m
نون	nwn	nu:n	نيون	nywn	niyo:n
همزة	hmzp	hamza	هنا	hnA	hina
هندسة	hndsp	handasa	هه	hh	hih
واربعة	w<rbEp	w?ArbA?\a	واربعتاشر	w<rbEtA$r	w?ArbA?\tA:SAr
واربعين	w<rbEyn	w?arbi?\i:n	واتناشر	w<tnA$r	w?itnA:SAr
واتنين	w<tnyn	w?itne:n	وانت	w<nt	winta
واحد	wAHd	wa:X\id	واحدة	wAHdp	waX\da
واخد	wAxd	wa:xid	وادي	wAdy	wa:di
واو	wAw	wA:w	وتاخد	wtAxd	wita:xud
وتسعة	wtsEp	wtis?\a	وتسعتاشر	wtsEtA$r	wtisa?\tA:SAr
وتسعمية	wtsEmyp	wtus?\umiyya	وتسعين	wtsEyn	wtis?\i:n
وتلاتاشر	wtlAtA$r	wtAlAttA:SAr	وتلاتة	wtlAtp	wtala:ta
وتلاتين	wtlAtyn	wtalati:n	وتلت	wtlt	wtilt
وتلتمية	wtltmyp	wtultumiyya	وتمانية	wtmAnyp	wtamanya

Table B.1 (continued)

Arabic	Buckwalter	SAMPA	Arabic	Buckwalter	SAMPA
وتمانين	wtmAnyn	wtamani:n	وتمنتاشر	wtmntA$r	wtAmAntA:SAr
وتمنمية	wtmnmyp	wtumnumiyya	وحداشر	wHdA$r	wX\idA:SAr
وخمسة	wxmsp	wxamsa	وخمستاشر	wxmstA$r	wxAmAstA:SAr
وخمسمية	wxmsmyp	wxumsumiyya	وخمسين	wxmsyn	wxamsi:n
ورايح	wrAyH	wrA:yiX\	وربع	wrbE	wrub?\
وربعمية	wrbEmyp	wrub?\umiyya	ورد	wrd	ward
ورقة	wrqp	wara?a	وسبعة	wsbEp	wsab?\a
وسبعتاشر	wsbEtA$r	wsAbA?\tA:SAr	وسبعمية	wsbEmyp	wsub?\umiyya
وسبعين	wsbEyn	wsab?\i:n	وستاشر	wstA$r	wsittA:SAr
وستة	wstp	wsitta	وستمية	wstmyp	wsuttumiyya
وستين	wstyn	wsitti:n	وسهلا	wshlA	wasahlan
وشه	w$h	wiSSu	وصول	wSwl	wus'u:l
وعشرة	wE$rp	w?\ASA:rA	وعشرين	wE$ryn	w?\iSri:n
ولد	wld	walad	ومية	wmyp	wmiyya
وميتين	wmytyn	wmite:n	ونص	wnS	wnus's'
وواحد	wwAHd	wwa:X\id	ياما	yAmA	yama
يديك	ydyk	yiddi:k	يسلمك	yslmk	ysallimak
يعني	yEny	ya?\ni	يغسل	ygsl	yiGsil
يله	ylh	yAllA	يناير	ynAyr	yana:yir
ينسون	ynswn	yansu:n	يه	yh	yih
يوفقك	ywfqk	ywaffa?ak	يوليه	ywlyh	yulya
يوليو	ywlyw	yulyu	يوم	ywm	yo:m
يونيه	ywnyh	yunya	يونيو	ywnyw	yunyu

Appendix C
Seinnheiser ME-3 Specifications

Seinnheiser ME-3 is a headset microphone of exceptional sound quality, the ME 3 is intended for music and speech applications. The super-cardioid condenser design offers excellent feedback rejection.

Table C.1 Specifications of the Seinnheiser ME-3 microphone

AF sensitivity	1.6 mV/Pa
Max. sound pressure level (active)	150 dB
Pick-up pattern	Super-cardioid
Transducer principle	Electret condenser

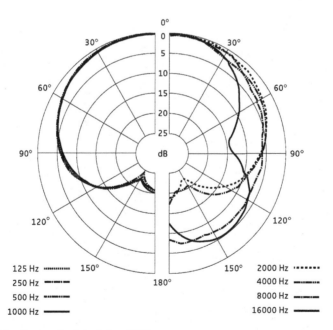

Fig. C.1 Polar diagram for Seinnheiser ME-3

M. Elmahdy et al., *Novel Techniques for Dialectal Arabic Speech Recognition*, DOI 10.1007/978-1-4614-1906-8, © Springer Science+Business Media New York 2012

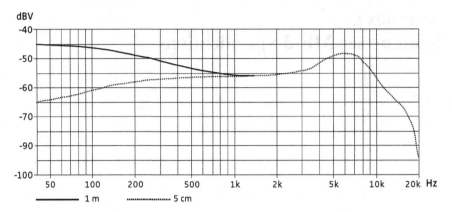

Fig. C.2 Frequency response curve for Seinnheiser ME-3

Appendix D
Buddy 6G USB Specifications

The Buddy 6G USB adapter manufacured by InSync speech Technologies, Inc. is based on the Micronas UAC3556b microchip. It has a built-in high-quality sound card, which replaces a desktop or laptop computer's sound card for high performance speech sound input and output. It offers full duplex operation for connection with microphone and speakers that is especially well suited to speech recognition applications.

Table D.1 Specifications of the Buddy 6G USB adapter based on the Micronas UAC3556b microship

Microphone	Supports 8/16 bit mono recording at 6.4 kHz to 48 kHz, sensitivity -54 ± 4 dB impedance < 650 Ohms.
Speaker Output	Supports 16/24 bit mono/stereo at 6.4 kHz to 48 kHz. Includes a low power stereo amplifier.
Signal to Noise Ratio	SNR is typically -92 dB for A/D (recording) and -96 dB for D/A (playback).
Total Harmonic Distortion	THD is better than -90 dB for both A/D (recording) and D/A (playback).
Power	Self powered from USB bus with less than 100 mA current at 5V DC.
Operating Temperature	Minimum -10°C (14°F), maximum 70°C (158°F).
Storage Temperature	Minimum -40°C (-40°F), maximum 75°C (167°F).

M. Elmahdy et al., *Novel Techniques for Dialectal Arabic Speech Recognition*,
DOI 10.1007/978-1-4614-1906-8, © Springer Science+Business Media New York 2012

References

Abdou, S., Hamid, S. E., Rashwan, M., Samir, M., Abd-Elhamid, O., Shahin, M., and Nazih, W. (2006) Computer Aided Pronunciation Learning System Using Speech Recognition Techniques. In Proceedings of International Conference on Speech and Language Processing INTERSPEECH

Afify, M., Nguyen, L., Xiang, B., Abdou, S., and Makhoul, J. (2005) Recent Progress in Arabic Broadcast News Transcription at BBN. In Proceedings of International Conference on Speech and Language Processing INTERSPEECH, Lisbon, Portugal, pp. 1637–1640

Afify, M., Sarikaya, R., Kuo, H. J., Besacier, L., and Gao, Y. (2006) On the use of morphological analysis for dialectal Arabic speech recognition. In Proceedings of International Conference on Speech and Language Processing INTERSPEECH, Pittsburgh, Pennsylvania, pp. 277–280

Alshalabi, R. (2005) Pattern-based Stemmer for Finding Arabic Roots. Information Technology Journal 4(1), pp. 38–43

Appen Pty Ltd, Sydney, Australia (2006a) Iraqi Arabic Conversational Telephone Speech. Linguistic Data Consortium, University of Pennsylvania, LDC Catalog No.: LDC2006S45

Appen Pty Ltd, Sydney, Australia (2006b) Gulf Arabic Conversational Telephone Speech. Linguistic Data Consortium, University of Pennsylvania, LDC Catalog No.: LDC2006S43

Appen Pty Ltd, Sydney, Australia (2007) Levantine Arabic Conversational Telephone Speech. Linguistic Data Consortium, University of Pennsylvania, LDC Catalog No.: LDC2007S01

Atiyya, M., Choukri, K., and Yaseen, K. (2005) Specifications of the Arabic Written Corpus. Nemlar project

Barras, C., Geoffroisb, E., Wuc, Z., and Libermanc, M. (2000) Transcriber: Development and use of a tool for assisting speech corpora production. Speech Communication 33(1–2), pp. 5–22

Billa, J., Noamany, M., Srivastava, A., Liu, D., Stone, R., Xu, J., Makhoul, J., and Kubala, F. (2002) Audio Indexing of Arabic Broadcast News. In Proceedings of the IEEE International Conference on Acoustics, Speech, and Signal Processing (ICASSP), vol. 1, pp. 5–8

Buckwalter, T. (2002a) Arabic Transliteration. URL: http://www.qamus.org/transliteration.htm

Buckwalter, T. (2002b) Buckwalter Arabic Morphological Analyzer Version 1.0. Linguistic Data Consortium, University of Pennsylvania, LDC Catalog No.: LDC2002L49

Cambridge (2010) HTK—Hidden Markov Model Toolkit—Speech Recognition toolkit. URL: http://htk.eng.cam.ac.uk/

Canavan, A., Zipperlen, G., and Graff, D. (1997) CALLHOME Egyptian Arabic Speech. Linguistic Data Consortium, University of Pennsylvania, LDC Catalog No.: LDC97S45

Clarkson, P., and Rosenfeld, R. (1997) Statistical Language Modeling Using the CMU-Cambridge Toolkit. In Proceedings of ISCA Eurospeech

Carnegie Mellon University (2010a) Sphinx—Speech Recognition Toolkit. URL: http://cmusphinx.sourceforge.net/

Carnegie Mellon University-Cambridge (2010b) CMU-Cambridge Statistical Language Modeling toolkit. URL: http://www.speech.cs.cmu.edu/SLM/toolkit.html

M. Elmahdy et al., *Novel Techniques for Dialectal Arabic Speech Recognition*, 103
DOI 10.1007/978-1-4614-1906-8, © Springer Science+Business Media New York 2012

Darwish, K. (2002) Building a shallow Arabic morphological analyzer in one day. In Proceedings of ACL workshop on computational approaches to semitic languages

Djoudi, M., Fohr, D., and Haton, J. P. (1989) Phonetic study for automatic recognition of Arabic. In Proceedings of first European conference on speech communication and technology (Eurospeech), Paris, France, pp. 2268–2271

Djoudi, M., Aouizerat, H., and Haton, J. P. (1990) Phonetic study and recognition of standard Arabic emphatic consonants. In Proceedings of First International conference on spoken language processing (ICSLP), Kobe, Japan, pp. 957–960

El-Halees, Y. (1989) A study of subglottal pressure for emphatic and non-emphatic sounds in Arabic. In Proceedings of first European conference on speech communication and technology (Eurospeech), Paris, France

Elmahdy, M., Gruhn, R., Minker, W., and Abdennadher, S. (2009a) Survey on Common Arabic Language Forms from a Speech Recognition Point of View. In Proceedings of the International Conference on Acoustics (NAG-DAGA), Rotterdam, Netherlands, pp. 63–66

Elmahdy, M., Gruhn, R., Minker, W., and Abdennadher, S. (2009b) Effect of Gaussian Densities and Amount of Training Data on Grapheme-Based Acoustic Modeling for Arabic. In Proceedings of the IEEE international conference on natural language processing and knowledge engineering (IEEE NLP-KE), Dalian, China

Elmahdy, M., Gruhn, R., Minker, W., and Abdennadher, S. (2009c) Modern Standard Arabic Based Multilingual Approach for Dialectal Arabic Speech Recognition. In International Symposium on Natural Language Processing (SNLP), Bangkok, Thailand, pp. 169–174

Elmahdy, M., Gruhn, R., Minker, W., and Abdennadher, S. (2010) Cross-Lingual Acoustic Modeling for Dialectal Arabic Speech Recognition. In Proceedings of International Conference on Speech and Language Processing INTERSPEECH, Makuhari, Japan, pp. 873–876

Elmahdy, M., Gruhn, R., Abdennadher, S., and Minker, W. (2011) Rapid Phonetic Transcription using Everyday Life Natural Chat Alphabet Orthography for Dialectal Arabic Speech Recognition. In Proceedings of the IEEE International Conference on Acoustics, Speech, and Signal Processing (ICASSP), Prague, Czech Republic

ELRA: European Language Resources Association (2010) URL: http://www.elra.info/

Fegen, C., Steker, S., Soltau, H., Metze, F., and Schultz, T. (2003) Efficient Handling of Multilingual Language Models. In Proceedings of Automatic Speech Recognition and Understanding Workshop (ASRU), St. Thomas, Virgin Islands, pp. 441–446

Ferguson, C. (1959) Diglossia. Word 15, pp. 325–340

Fung, P., and Schultz, T. (2008) Multilingual Spoken Language Processing. IEEE Speech Processing Magazine 25(3), pp. 89–97

Gal, Y. (2002) An HMM approach to vowel restoration in Arabic and Hebrew. In Proceedings of the ACL-02 workshop on Computational approaches to semitic languages, USA, Association for Computational Linguistics

Gales, M. J. F., Diehl, F., Raut, C. K., Tomalin, M., Woodland, P. C., and Yu, K. (2007) Development of a Phonetic System for Large Vocabulary Arabic Speech Recognition. In Proceedings of Automatic Speech Recognition and Understanding Workshop (ASRU), Kyoto, Japan, pp. 24–29

Gibbon, D., Moore, R., and Winski, R. (1997) SAMPA computer readable phonetic alphabet. In Handbook of Standards and Resources for Spoken Language Systems. Mouton de Gruyter, Berlin. Part IV, section B

Google Labs (2009) Google Transliteration. URL: http://www.google.com/ta3reeb/

Google Labs (2010) Google Tashkeel. URL: http://tashkeel.google.com

Gruhn, R., and Nakamura, S. (2001) Multilingual, Speech Recognition with the CALLHOME Corpus (ASJ2001), vol. 1. Acoustical Society of Japan, Japan, pp. 153–154

Gu, L., Zhang, W., Tahir, L., and Gao, Y. (2007) Statistical Vowelization of Arabic Text for Speech Synthesis in Speech-to-Speech Translation Systems. In International Conference on Speech and Language Processing INTERSPEECH, Antwerp, Belgium, pp. 1901–1904

Habash, N., and Rambow, O. (2007) Arabic Diacritization through Full Morphological Tagging. In Proceedings of NAACL HLT, pp. 53–56

Hassan, Z. M., and Esling, J. H. (2007) Laryngoscopic (Articulatory) and Acoustic Evidence of a Prevailing Emphatic Feature Over the Word in Arabic. In Proceedings of the 16th International Congress of Phonetic Sciences

Hinds, M., and Badawi, E. (2009) A Dictionary of Egyptian Arabic. Librairie du Liban, Reprinted

Holes, C. (2004) Modern Arabic: Structures, Functions, and Varieties. Georgetown University Press, Washington

Huang, X., Acero, A., and Hon, H. (2001) Spoken language processing: a guide to theory, algorithm, and system development. Prentice Hall, New York

ISO 8859-6 (1987) Information processing—8-bit single-byte coded graphic character sets—Part 6: Latin/Arabic alphabet. International Organization for Standardization

Jurafsky, D., and Martin, J. H. (2009) Speech and language processing: An introduction to natural language processing, computational linguistics, and speech recognition, second edition. Prentice Hall, New York

Kaye, A. S. (1970) Modern Standard Arabic and the Colloquials. Lingua 24, pp. 374–391

Kilany, H., Gadalla, H., Arram, H., Yacoub, A., El-Habashi, A., and McLemore, C. (2002) Egyptian Colloquial Arabic Lexicon. Linguistic Data Consortium, University of Pennsylvania, LDC Catalog No.: LDC99L22

Kirchhoff, K., and Vergyri, D. (2005) Cross-Dialectal Data Sharing For Acoustic Modeling in Arabic Speech Recognition. Speech Communication 46(1), pp. 37–51

Kirchhoff, K., Bilmes, J., Das, S., Duta, N., Egan, M., Ji, G., He, F., Henderson, J., Liu, D., Noamany, M., Schone, P., Schwarta, R., and Vergyri, D. (2002) Novel approaches to Arabic speech recognition: report from the 2002 Johns-Hopkins summer workshop. Technical report, Johns Hopkins University

Lagally, K. (1992) ArabTEX Typesetting Arabic with vowels and ligatures. In Proceedings of the EuroTEX92 conference, Prague

Lamel, L., Messaoudi, A., and Gauvain, J. (2007) Improved Acoustic Modeling for Transcribing Arabic Broadcast Data. In International Conference on Speech and Language Processing INTERSPEECH, pp. 2077–2080

Lamere, P., Kwok, P., Gouvea, E. B., Raj, B., Singh, R., Walker, W., and Wolf, P. (2003) The CMU SPHINX-4 speech recognition system. In Proceedings of the IEEE International Conference on Acoustics, Speech, and Signal Processing (ICASSP), vol. 46(1), pp. 37–51

Linguistic Data Consortium (LDC) (2010) University of Pennsylvania. URL: http://www.ldc.upenn.edu/

Lee, C., and Gauvain, J. (1993) Speaker Adaptation Based on MAP Estimation of HMM Parameters. In Proceedings of the IEEE International Conference on Acoustics, Speech, and Signal Processing (ICASSP), pp. II–558

Leggetter, C. J., and Woodland, P. C. (1995) Maximum likelihood linear regression for speaker adaptation of the parameters of continuous density hidden Markov models. Computer Speech and Language 9, pp. 171–185

Maamouri, M., Graff, D., Jin, H., Cieri, C., and Buckwalter, T. (2004) Dialectal Arabic Orthography-based Transcription and CTS Levantine Arabic Collection. Paper presented at the Parallel STT-NA Tracks Session of the EARS RT-04 Workshop, Palisades IBM Executive Center, New York

Maamouri, M., Graff, D., and Cieri, C. (2006) Arabic Broadcast News Transcripts. Linguistic Data Consortium, University of Pennsylvania, LDC Catalog No.: LDC99L22

Maamouri, M., Buckwalter, T., Graff, D., and Jin, H. (2007) Fisher Levantine Arabic Conversational Telephone Speech. Linguistic Data Consortium, University of Pennsylvania, LDC Catalog No.: LDC2007S02

Maegaard, B., Damsgaard, J. L., Krauwer, S., and Choukri, K. (2004) NEMLAR: Arabic Language Resources and Tools. In Proceedings of Arabic Language Resources and Tools Conference, Cairo, Egypt, pp. 42–54

Makhoul, J., Zawaydeh, B., Choi, F., and Stallard, D. (2005) BBN/AUB DARPA Babylon Levantine Arabic Speech and Transcripts. Linguistic Data Consortium, University of Pennsylvania, LDC Catalog No.: LDC2005S08

Messaoudi, A., Lamel, L., and Gauvain, J. (2004) Transcription of Arabic Broadcast News. In International Conference on Spoken Language Processing (INTERSPEECH), Jeju Island, Korea, pp. 1701–1704

Messaoudi, A., Gauvain, J., and Lamel, L. (2006) Arabic Broadcast News Transcription using a One Million Word Vocalized Vocabulary. In Proceedings of the IEEE International Conference on Acoustics, Speech, and Signal Processing (ICASSP), vol. 1, pp. 1093–1096

Microsoft Innovation Lab, Cairo (2009) Microsoft Maren. URL: http://www.microsoft.com/middleeast/egypt/cmic/maren/

Nelken, R., and Shieber, S. M. (2005) Arabic Diacritization Using Weighted Finite-State Transducers. Workshop On Computational Approaches To Semitic Languages 5(2), pp. 79–86

Newman, D. (2002) The Phonetic Status of Arabic within the World's Languages: The Uniqueness of the Lughat Al-Aaad. Antwerp papers in linguistics 100, pp. 65–75

Ney, H., Essen, U., and Kneser, R. (1994) On structuring probabilistic dependencies in stochastic language modeling. Computer Speech and Language 8(1), pp. 1–28

Ng, T., Nguyen, K., Zbib, R., and Nguyen, L. (2009) Improved Morphological Decomposition for Arabic Broadcast News Transcription. In Proceedings of the IEEE International Conference on Acoustics, Speech, and Signal Processing (ICASSP), Taipei, Taiwan, pp. 4309–4311

Paulsson, K., Choukri, K., Mostefa, D., DiPersio, D., Glenn, M., and Strassel, S. (2009) A Large Arabic Broadcast News Speech Data Collection. In Proceedings of the Second International Conference on Arabic Language Resources and Tools, Egypt, pp. 280–284

Rabiner, L., and Juang, B. (1993) Fundamentals of Speech Recognition. Prentice Hall, New York

Rabiner, L. R. (1989) A Tutorial on Hidden Markov Models and Selected Applications in Speech Recognition. Proceedings of the IEEE 77(2), pp. 257–286

Razak, Z., Ibrahim, N. J., Idris, M. Y. I., Tamil, E. M., Yakub, M., Yusoff, Z. M., and Rahman, N. N. A. (2008) Quranic Verse Recitation Recognition Module for Support in j-QAF Learning: A Review. IJCSNS International Journal of Computer Science and Network Security 8(8), pp. 207–216

RDI (2007) Fassieh. URL: http://www.rdi-eg.com/

Rybach, D., Hahn, S., Gollan, C., Schluter, R., and Ney, H. (2007) Advances in Arabic Broadcasr News Transcription At RWTH. In Proceedings of Automatic Speech Recognition and Understanding Workshop (ASRU), Kyoto, Japan, pp. 449–454

The Nemlar project (2005) URL: http://www.nemlar.org/

Sarikaya, R., Emam, O., Zitouni, I., and Gao, Y. (2006) Maximum Entropy Modeling for Diacritization of Arabic Text. In Proceedings of International Conference on Speech and Language Processing INTERSPEECH, pp. 145–148

Schultz, T., and Waibel, A. (2001) Language Independent and Language Adaptive Acoustic Modeling for Speech Recognition. Speech Communication 35, pp. 31–51

Stevens, V., and Salib, M. (2005) A Pocket Dictionary of the Spoken Arabic of Cairo. The American University in Cairo Press, Cairo

Vergyri, D., and Kirchhoff, K. (2004) Automatic diacritization of Arabic for acoustic modeling in speech recognition. In Proceedings of COLING Computational Approaches to Arabic Script-based Languages, Geneva, Switzerland, pp. 66–73

Vergyri, D., Kirchhoff, K., Gadde, R., Stolcke, A., and Zheng, J. (2005) Development of a conversational telephone speech recognizer for Levantine Arabic. In Proceedings of International Conference on Speech and Language Processing INTERSPEECH, Lisboa, pp. 1613–1616

Waibel, A., Geutner, P., Mayfield-Tomokiyo, L., Schultz, T., and Woszczyna, M. (2000) Multilinguality in Speech and Spoken Language Systems. Proceedings of the IEEE, Special Issue on Spoken Language Processing 88(8), pp. 1297–1313

Xiang, B., Nguyen, K., Nguyen, L., Schwartz, R., and Makhoul, J. (2006) Morphological decomposition for Arabic broadcast news transcription. In Proceedings of the IEEE International Conference on Acoustics, Speech, and Signal Processing (ICASSP), vol. I, pp. 1089–1092

Yaghan, M. A. (2008) Arabizi: a contemporary style of Arabic slang. Design Issues 24(2), pp. 39–52

Yaseen, M., Attia, M., Maegaard, B., Choukri, K., Paulsson, N., Haamid, S., Krauwer, S., Bendahman, C., Fersoe, H., Rashwan, M., Haddad, B., Mukbel, C., Mouradi, A., Al-Kufaishi, A., Shahin, M., Chenfour, N., and Ragheb, A. (2006) Building Annotated Written and Spoken Arabic LRs in NEMLAR Project. In Proceedings of International Conference on Language Resources and Evaluation (LREC)

Young, S., Evermann, G., Gales, M., Hain, T., Kershaw, D., Liu, X., Moore, G., Odell, J., Ollason, D., Povey, D., Valtchev, V., and Woodland, P. (1996) The HTK Book. Cambridge University Press, Cambridge

Zitouni, I., Olive, J., Iskra, D., Choukri, K., Emam, O., Gedge, O., Maragoudakis, E., Tropf, H., Moreno, A., Rodriguez, A. N., Heuft, B., and Siemund, R. (2002) ORIENTEL: Speech-Based Interactive Communication applications for the Mediterranean and the Middle East. In Proceedings of International Conference on Speech and Language Processing INTERSPEECH, pp. 325–328

Index

M. Elmahdy et al., *Novel Techniques for Dialectal Arabic Speech Recognition*,
DOI 10.1007/978-1-4614-1906-8, © Springer Science+Business Media New York 2012